T0357846

The Cost of Conviction

How Our Deepest Values Lead Us Astray

Steven Sloman

The MIT Press
Cambridge, Massachusetts
London, England

The MIT Press
Massachusetts Institute of Technology
77 Massachusetts Avenue, Cambridge, MA 02139
mitpress.mit.edu

The MIT Press would like to thank the anonymous peer reviewers who provided comments on drafts of this book. The generous work of academic experts is essential for establishing the authority and quality of our publications. We acknowledge with gratitude the contributions of these otherwise uncredited readers.

This book was set in ITC Stone Serif Std and ITC Stone Sans Std by New Best-set Typesetters Ltd. Printed and bound in the United States of America.

Library of Congress Cataloging-in-Publication Data

Names: Sloman, Steven, author.
Title: The cost of conviction : how our deepest values lead us astray / Steven Sloman.
Description: Cambridge, Massachusetts : The MIT Press, [2025] | Includes bibliographical references and index.
Identifiers: LCCN 2024044336 (print) | LCCN 2024044337 (ebook) | ISBN 9780262049825 (hardcover) | ISBN 9780262383820 (pdf) | ISBN 9780262383837 (epub)
Subjects: LCSH: Values. | Decision making. | Group identity.
Classification: LCC BF778 .S56 2025 (print) | LCC BF778 (ebook) | DDC 158—dc23/eng/20250203
LC record available at https://lccn.loc.gov/2024044336
LC ebook record available at https://lccn.loc.gov/2024044337

10 9 8 7 6 5 4 3 2 1

EU product safety and compliance information contact is: mitp-eu-gpsr@mit.edu

To my father, Leon Sloman (1929–2024), who entrusted to me his most sacred values.

Contents

1 What This Book Is About

Zealotry is not new. Indeed, the original Zealots were a young, armed, extremist group rebelling against the Roman occupation of Judaea two millennia ago. They were resolved to fight off the Roman forces led by Vespasian in the first century CE—later to become emperor of Rome—and then by his son Titus. Despite losing, remnants of the Zealots refused to surrender. They escaped the burning temple in Jerusalem and were forced to flee into the Upper City of Jerusalem and fortresses along the Jordan River. They vainly tried to defend their land from the Romans.[1]

The Zealots valued freedom so much that they risked everything they had including their lives. On some accounts, some of them even engaged in violent, murderous attacks on their Hebrew brethren who they accused of colluding with the Roman authorities. Their longest and bravest fight ended with mass suicide on Masada, a desert fortress on the western shore of the Dead Sea that still serves as a major tourist attraction in Israel. Rather than surrendering to the Romans and betraying their sacred values, they paid the ultimate price. Thousands of years later, we still use the word "zealot" to describe someone with extreme, uncompromising positions.

Radical extremists are often treated as heroes, especially when we are aligned with their principles. That is why the French made the warrior Joan of Arc a patron saint; she held her direct relationship with God sacred and was willing to be burned at the stake for it at the age of nineteen. It is heroic to fight for one's values and beliefs when facing the possibility of death. But it is not always recommended. Too often we give ourselves the choice of presenting as zealots in the fear that if we don't, we will be seen as lacking principle. We feel we have to take an extreme, uncompromising position or we will be seen as devoid of moral courage, or worse, the embodiment of evil. We are either on board with our community's fundamentalist values or are told that our values deserve to be crushed by the sledgehammer of justice. In this book, I will show you when and why uncompromising positions based on sacred values have merit, and also outline an alternative way to present yourself and make decisions—a way that lowers the temperature and encourages more thoughtfulness. This alternative is called consequentialism, making decisions based not on sacred values but instead on the outcomes associated with the various options.

The death-defying heroics of zealotry are an extreme form of a facet of pretty much all humans. Almost everyone holds certain actions as sacred. Show gratitude when someone has done you a favor, even if you will never encounter that person again. Tell the truth under all but extraordinary circumstances. Give to worthy causes, and some people take bonus credit if they do so anonymously. Sometimes it's nonactions that we hold sacred. Don't swear in front of someone who might be offended by it. Don't step on baby rabbits. And then of course there are sacred values shared by some that have a religious origin. Treat thy neighbor as you would have them treat you. Don't say God's name in vain. Don't eat pork.

These prescriptions and proscriptions do not all have exactly the same character. But they do have a critical feature in common. Notice that they all start with a verb (show, tell, give, and don't swear, step, say, or eat). That's because they concern action, what to do. Values that are not sacred do not; they concern consequences. For instance, you might value being able to provide for your family. That's not an action; it's an outcome. It doesn't imply doing anything specific or not doing anything specific. It just says that whatever you do, you'd prefer to end up with a reasonable amount of money. In contrast, a sacred value like "don't eat pork" has specific implications for action: Do whatever you want as long as you're not putting pork into your mouth.

Most of the discussion in this book will focus on the more prosaic decisions we make every day. It will also try to explain how decision making by sacred values can get out of control. It can lead to a cycle that drives communities apart, leading on occasion to the violence of zealotry.

Evidence for Two Kinds of Values

If you're not completely convinced that sacred values about action and consequentialist values about outcomes are different in important ways, consider the following vignettes inspired by Philip Tetlock, a psychologist at the University of Pennsylvania, and his colleagues.[2]

Vignette 1: You only have $50 available. A neighbor who just lost their job comes by and says they need $50 to buy a meal plan for their child, otherwise their child will not have lunch at school. Can you give them the money? You ask two friends for their advice. Frankie thinks about it for quite a while and then advises you to give your neighbor the money. Charlie hardly thinks at all, advising you immediately to give the neighbor the money. On the basis of their advice, who would you say is more morally praiseworthy, Frankie or Charlie?

Vignette 2: You only have $50 available. A neighbor who just lost their job comes by and says they need $50 to buy a meal plan for their child, otherwise their child will not have lunch at school. Can you give them the money? But there's a problem. You just got a text from an old schoolmate asking to borrow $50 to buy the medication they need to cure a serious illness. You ask two friends for their advice. Frankie thinks about it for quite a while and then advises you to give your neighbor the money. Charlie hardly thinks at all, advising you immediately to give the neighbor the money. On the basis of their advice, who would you say is more morally praiseworthy, Frankie or Charlie?

Most people think Charlie is more praiseworthy in vignette 1, but Frankie is more praiseworthy in vignette 2. Vignette 1 asks about a "taboo" trade-off between a sacred value (helping a child) and money. A moral person should have no trouble choosing the sacred value, so Charlie's lack of thought and immediate response is more praiseworthy. Vignette 2 involves a "tragic" trade-off between two sacred values (helping a child versus helping a sick schoolmate). Morally speaking, this should be hard because it forces you to violate a sacred value by not giving to someone, and that hurts. Hard decisions take time, so Frankie, who deliberates before answering, is showing more evidence of principled behavior in this case.

The intent of this experiment is to pump your intuition that sacred values are different from consequentialist ones about outcomes like money. Sacred values overwhelm material desires without even having to think about it. But having to choose between sacred values puts us between a rock and a hard place because such values should never be violated. As we will see, violating them produces feelings of contamination. When somebody else violates them, we feel outrage.

Conflicting Frames

This distinction between sacred values that are about action as opposed to values about consequences identifies two very different ways of framing what you like, what you want, and what you consider right or wrong. As the taboo trade-off example shows, these frames can be at odds with each other. That slow-roasted pork sandwich might look delicious even if it is forbidden to me (or because it is forbidden to me!). If so, the tasty outcome of biting into a mouthwatering, flavorful delicacy is at odds with my religious or moral decision to avoid certain kinds of food. On a larger scale, I might get an opportunity to take a lucrative position at a consulting firm. The position will pay me enough to send my kids to college, but the firm is guilty of providing advice to a tobacco company about how to sell cigarettes to young kids. In this case, I have to decide whether I am willing to accept substantial payment for violating what is for many people a sacred value—not inducing children to smoke. Should I take the job?

Such decisions force one to choose between a taboo option—violating a sacred value—and a consequence—material gain. But rarely are choices this stark, and even identifying which considerations concern sacred values and which concern consequences can be difficult. Most decisions have multiple dimensions, with some of them involving material gain, some of them touching on values we hold dear, and some involving both. Perhaps limiting your impact on the environment is important to you—a sacred value. If you are buying a car, you might have to trade off how much your purchase affects the environment with the other features that matter to you (e.g., size, reliability, and speed). The simplest way to do this is to consider how much the car will impact the environment, framing the impact on the environment as a consequence as opposed to a sacred value. That way, you can address a question such as whether it is better to get an electric car by asking, for instance, whether the electricity in your region is generated by renewables or fossil fuels. It's likely that you'll end up making some trade-offs in favor of nonenvironmental factors because the most environmentally friendly car is probably not the fastest, safest, and most reliable one. Limiting yourself to your sacred value just won't do. My main goal in writing this book is to convince you, dear reader, that even though we cannot be solid members of society without deploying both sacred values and an analysis of consequences to make decisions, many social problems arise because we have a bad habit of relying more on sacred values than we should. Trade-offs are almost always necessary.

To be clear, most of the time this is obvious. Our mundane, everyday decisions tend to be governed by their consequences. We have coffee in the morning because we feel tired and lethargic without it, and probably like the effect it has on our taste buds. We choose coffee because its consequences have rewarded us in the past. This applies to much of our day: working, exercising, eating, and relaxing. Much of what we do, we do to achieve positive consequences and avoid negative ones.

But even everyday activities can be driven by sacred values. Our work may involve helping others, eating may involve avoiding animal flesh, and relaxing might involve prayer. Sacred values influence us all the time, as they should. Having at least some sacred values is an essential property of a good citizen because morality requires us to hold certain values sacred (like "avoid harming others"). But sacred values have a dark side. For one, they tend to be inflexible and unaccommodating to other values as well as the contingencies that inevitably arise in life. Pacifists may hold avoiding war to be sacred, but that can conflict with a duty to protect one's family when under attack. The conflict produced by such tragic trade-offs can lead to enormous tension within the individual who faces the moral dilemma of remaining true to their values or watching a loved one get hurt. It can also lead to enormous tension in society when different groups holding different values are unwilling to compromise. Such differences are currently tearing apart the United States, as those who hold blue state values often believe that those with red state values are destroying democracy and limiting freedom arbitrarily, while red staters believe exactly the same thing about those with blue state values. A similar dynamic is occurring elsewhere in the world.

Being good does not always require putting one's sacred values on display. Most issues can be framed in terms of the consequences they produce. Rather than asking whether it is right or wrong to, say, allow people to carry concealed weapons, we can ask what the consequences of such a policy would be. And asserting a consequentialist frame makes conversation possible. It turns out that people on both sides desire many of the same consequences (lower murder and suicide rates, and fewer mass shootings). By framing issues in terms of how to achieve consequences, we can avoid violating people's sacred values and find common ground. But there is a cost: Talking about consequences is hard, and it requires facts and expertise. We are all severely limited in our ability to discuss consequences by our limited knowledge.

Why I Wrote This Book

We could just admit our ignorance and appeal to experts. But we frequently don't for a reason elaborated at length in my last book, *The Knowledge Illusion: Why We Never Think Alone*, written with Phil Fernbach. We showed that people tend to think they understand how things work better than they do. That's the knowledge illusion, and it applies to almost everything, from our understanding of ballpoint pens and toilets, to political policies and mental health challenges.[3] We do not admit our ignorance to ourselves. Since that book was written, I have come to realize that one reason we overestimate our understanding about social issues is because our gut reaction is to think about them in terms of sacred values, and this makes them seem simple and understandable. If I mention health care policy, your first reaction may be to think about a sacred value, that health care should be universal, or, alternatively, that government should stay out of the way and allow a free health care market. Neither slogan actually captures very much about health care as it is an exceedingly complex topic, the subject of countless studies, books, and debates. I can convince you of this simply by asking you to explain how health care policies work. What have been the actual detailed consequences of Obamacare? This forces you to frame the issue consequentially, in terms of how policies lead to outcomes. Those who take this request seriously discover that they cannot explain very much because there are an overwhelming number of details to know, and this lowers their sense of understanding.

The Knowledge Illusion also argues that our inflated sense of understanding occurs because we confuse what others know with what we know. Specific members of our community know a lot about stuff—including issues like health care—and generally, with a little work, we can find out what they know. The fact that we can access information that is sitting in other people's heads fools us into thinking we know it ourselves. After all, we evolved in tribes. Thinking, reasoning, and decision making are all functions that benefit by considering multiple perspectives, taking advantage of many people's knowledge. So our ancestors thought collectively. We still do; we never think alone. Our sacred values are handed down by our communities (thou shalt not murder), and our beliefs about how things work are distributed across the community (plumbers fix toilets, health care specialists help people and inform policy, etc.).

Thus we appeal to experts unconsciously by thinking we know things that only experts actually know. As a result, we often fail to appeal to experts when we need to because that requires humility and trust. So we do the easy thing: We retreat to our sacred values and make pronouncements about what's right and wrong that channel our community's position. Politicians do this all the time. Ask them to spell out the consequences of their policy proposal and what you will get back is the sacred value used to support it, with the implied hope that you share that value. This tactic was deployed masterfully by Theresa May, UK prime minister from 2016 to 2019. She was asked by a BBC interviewer about grammar schools, publicly funded schools that some accuse of being elitist, "Are you prepared to allow more grammar schools to open and existing grammar schools to expand?" Her response after some dithering was, "I want to make sure that children have those opportunities, that all schools are offering a good education to children."[4] An appeal to sacred values can be an effective way to dodge a question.

The burden of this book will be to show that our psychological habits are a major force in today's culture wars and polarization. We lean too much on our sacred values, and would do better by spending more time thinking about and discussing consequences. I will show, for instance, that framing issues in terms of sacred values makes people less willing to compromise than framing them in terms of consequences.

My other agenda in writing this book is continuing what I've been doing for forty years: trying to figure out how the mind works, how people think. A lot of theories in cognitive science—the field that studies thought—assume that people are consequentialist, that we are all busy trying to get the best outcomes. We will encounter some of these theories as we go. We will also encounter psychological theories that assume that people are guided by rules that apply to action and not to consequences. We might avoid someone, for instance, because we feel disgusted by them after discovering they hold beliefs that we find deplorable, even though we know that talking to them would likely ameliorate the hard feelings. On these theories, thinking—if there is any—is all about which action to take, not what the outcome will be. There are also more sophisticated theories that say we do both and have distinct cognitive systems for thinking about consequences and what causes them as well as thinking about rules for action.[5]

To put my cards on the table, I also believe we have distinct cognitive systems for processing different kinds of information, but I do not believe those systems are specialized for sacred values versus consequentialist reasoning. I think there is overwhelming evidence that distinct cognitive processes generate responses intuitively (which US president freed the slaves?) as opposed to generating responses with deliberation (which US president succeeded George Washington?). But both sacred values and consequentialist reasoning involve intuition and deliberation. Sacred values reasoning sometimes involves an intuitive reaction that provokes disgust or anger, and sometimes it involves careful thought about which sacred value is most relevant at a given time (e.g., social justice or freedom of speech). Similarly, consequential reasoning can be intuitive (I'll take the marshmallow–peanut butter flavor because it tastes better) and sometimes it involves deliberation (what kind of sanitation system should I have installed in my new house?). Sure, each kind of reasoning may engage particular cognitive systems more than others, and each cognitive system may tend to be more active in one kind of reasoning, but all cognitive systems are involved in both. Understanding how people process the two kinds of information is important in several ways. For one, it helps us figure out the best way to influence someone. It turns out that we cannot steer people toward sacred value versus consequentialist arguments simply by making it easier or harder to deliberate.

Scope of the Book

This book concerns how individuals as well as organizations, communities, and other groups decide what to do. It will provide a lot of background on what we know about how people make decisions and review ideas about how we should make decisions. It should serve as a primer on the psychology of decision making and its implications for how to make decisions. Its underlying theme will be that sacred values have their place, but that we overindulge in them.

Most of what we do, most of the choices we make, are made without thinking. They are made out of habit or by processes that are unconscious.

There is lots of great research about aspects of decision making that do not involve thought—about unconscious motivation, habit, addiction, emotion, and unconscious processing. My coverage of these topics will be narrow, but I will address these influences on action because they influence decision making at every stage.

What is special about thinking is that it provides reasons for our choices and actions, reasons that we can deliberate about either alone or with others. More often than we think, our deliberations involve others.[6] Sometimes we communicate directly with others. Even more frequently, others influence us indirectly: We make use of ideas that are sitting in other people's heads. For instance, we let experts make our decisions for us as we do when we let an accountant (or computer program) fill out our tax forms. Whether collaborating with or outsourcing to others, thinking is a group activity. Social welfare tends to be decided collectively, based on reasons that are conscious and articulable.

The study of thought was the domain of philosophy prior to the modern era. Now it is studied more by psychologists and cognitive scientists. Our subject, the dual roles of sacred and outcome values in decision making, has its origin in the philosophy of ethics. Philosophers have long distinguished bases for making moral judgments. One is closely related to sacred values and is frequently called deontology. The great philosopher Immanuel Kant (1724–1804) is often thought of as the originator of deontology, but there are deontological ideas that long predate Kant. One is the golden rule, referring to Jesus's advice to do unto others as you would have them do unto you.[7] This idea is shared in some form by most of the world's religions.

The other basis for moral judgment relates to deciding based on outcomes and is what we mean by "consequentialism." It includes today's most common standard for evaluating decisions, expected utility theory, and is often traced back to philosopher Jeremy Bentham (1748–1832). Some philosophers also appeal to a third basis for moral judgment, virtue ethics, a standard typically ascribed to Aristotle, but it isn't as general as the other two, and I will discuss it only in passing.

While philosophers remain the standard-bearers of the question, "How should we act?," it is psychologists who have made the most progress answering the question, "How do we act?" Sometimes we act consequentially. We determine what we want to achieve and then figure out the best way to do so. But frequently—more often than we think—we base our actions on sacred values or at least on rules of behavior that are reminiscent of sacred values.

Sometimes we use sacred values strategically. Consider Terry, a political candidate who wants to be elected to Congress and is faced with two opponents. One is an idealistic progressive who holds equality sacred, and argues that everybody should have health insurance and a guaranteed income. The other is a firebrand conservative who values individual freedom, chafing against gun control laws and the redistribution of wealth. In other words, the campaigns of Terry's competition are focused on their treasured (though competing) sacred values, not on detailing how to achieve outcomes. Terry is aiming for the middle. Terry still believes, perhaps naively, that there are enough moderates out there who will vote for a middle-of-the-road agenda. What should Terry do? Should Terry make a case for pragmatism and focus on how best to achieve results? Or should Terry develop yet another set of sacred values—perhaps a compromise position—and promote that?

On the one hand, Terry could take a consequentialist approach by focusing on outcomes that are almost universally favored: a stronger economy, higher wages, less crime, less pollution, better roads, and so on. Who doesn't value these things? Then Terry could talk about how best to achieve these goals: how government bureaucracy works, incentive structures that could be put in place for the various interest groups, and how to fund each step of the process. In short, Terry could present as a policy wonk: detail oriented, knowledgeable, and hardworking. No doubt Terry would attract some votes this way. Maybe your vote. But the data show that this strategy only appeals to a select few and fails to "excite the base" (a common refrain in the news these days). When faced with a choice between candidates talking from the heart, espousing strongly held values that touch on ideologically central topics, versus a candidate giving a policy lecture whose ins and outs are unfamiliar and hard to follow, the ideologue has a distinct advantage. Just ask Hillary Clinton.

On the other hand, Terry could present as an ideologue like his opponents. Terry could choose a mix of the opponents' sacred values to create a campaign around. They might try to appeal to equality by supporting everyone's right to medical insurance, capturing some of the left-wing competitor's supporters, but also take a stand for freedom by arguing against strong gun controls like the right-wing competitor. Then Terry could argue for a position that doesn't align with any extremists' agenda and push the sacred value of heterodoxy—that is, that good ideas can come from everywhere. The beauty of that approach is that Terry can come across as a real person with human emotions, a sense of outrage, and strong values, and not a boring policy wonk.

What should Terry do? You decide. Each approach has merit, as we will see in the rest of this book.

There is an irony here. Let's say that Terry takes the sacred value road (it's hard not to these days). The main motivation to do so is to achieve the best outcome, getting elected. So Terry's appeal to sacred values is motivated by a consequentialist analysis. Terry is being deeply pragmatic—so pragmatic that the campaign will pretend that their motivation is not pragmatic but rather based on deep-seated sacred values. One possible consequence of going this route is that Terry might become convinced of the sacred values they are espousing themselves.

One lesson of this book is this: A moral person necessarily has at least some sacred values. To be an acceptable member of any community—except a community of ruthless consequentialists—a person must display certain sacred values defined by the norms of that community. And doing so can reap large benefits. It can reinforce social ties and help one build a coalition. It can excite passions, and is necessary to elicit outrage for just causes and incite action.

But humans overindulge in their sacred values. Sacred values make important social and personal issues seem simpler than they are. That apparent simplicity facilitates conversation, but gives people a false sense of understanding. For individuals and society to progress, people need to admit the full complexity and difficulty of the world we make decisions in.

This book will take two deep dives. In chapters 2 and 3, I will explore sacred values along with their philosophical and psychological bases. I'll reveal why they are essential (because they provide common ground, community, and identity). Then I'll look at why they are problematic (because they oversimplify complex issues, and make people intransigent and more extreme than they should be). Next, chapters 4–9 will introduce the main alternative to making decisions by sacred values, consequentialism. Consequentialism is a rich topic, with lively historical roots and applications to everything we do. It is the closest we have to a general theory of how to behave, and it has been championed by philosophers, economists, and cognitive scientists. And despite having taken some powerful punches, it is still standing. The book's focus is on how we think when making decisions, but much of the study of decision making concerns processes that we are not aware of. These processes, and how they relate to sacred values and consequentialism, are discussed in chapter 10. Finally, in chapters 11 and 12, I will pit sacred values and consequentialism against one another, showing where and when these different positions are viable and preferable, before drawing some lessons about their applicability to everyday life and society.

2 Why Sacred Values Are Essential

Imagine coming across someone who insists on wearing nothing but a fig leaf. Regardless of where they put the fig leaf, body parts will be revealed. Even a large fig leaf does not afford complete modesty. Imagine that is all this individual wears whether they are at a restaurant, at the opera, or doing yoga. This person is not made of stone and on a pedestal in a museum. They are a live flesh-and-blood human being. Would you befriend such a person?

Flouting social norms like this is not the best way to win friends and influence people. There are certainly venues where it would be acceptable—summer art festivals in the desert perhaps—but few restaurants, even fewer opera houses, and only very private yoga studios would tolerate it. Does behaving this way constitute a violation of a sacred value? To answer this question, we need to have a clearer understanding of what we mean by "sacred value."

Defining Sacred Values

The concept of a sacred value is related to a number of other concepts like religious code, moral principle, social convention, and social norm. What is the place of sacred values in this cluster? The term "sacred value" has its origins in religion. The word "religious" is almost equivalent to one sense of "sacred."[1] Religions have sacred books, sacred figures, and sacred beliefs. "Sacred values" seems to point to a related category whose sanctity should not be violated without eliciting outrage and condemnation.[2] Violations can come with punishment, sometimes harsh punishment (burning at the stake comes to mind). More on this in a moment.

These days psychologists use the term "sacred values" without the religious connotation. After all, many people are not religious yet they still hold strong principles to regulate their behavior. In the United States, for instance, fewer and fewer people claim links to organized religion. In 1937, 73 percent of Americans answered "yes" to the question, "Do you happen to be a member of a church, synagogue, mosque, or temple?" In 2021, only 47 percent did.[3] Yet people still have values about what actions are acceptable. You don't need a relationship with a priest, rabbi, imam, or other religious leader to believe that murder is wrong or that everyone has a right to liberty. Some theorists have jettisoned the term "sacred" and refer to these as "protected values" to avoid the religious flavor. But I'm going to stick to the term "sacred values" until I am convinced that religious and non-religious sacred values differ in some fundamental way other than having their source in organized religion. The term "sacred" strikes me as having the right connotations.

Let's get back to the question of definition. I tend to agree with philosopher of mind Jerry Fodor, who asserted, "I don't think that theoretical terms usually have definitions (for that matter, I don't think that non-theoretical terms usually do either)."[4] That is, I am not optimistic that I will be able to give you a set of necessary and sufficient conditions that separate sacred values from everything else. Definitions work in math and logic, but not in natural human discourse. Nevertheless, we can hone down the meaning of the term "sacred values" to be quite precise.

Start with "values." To say you value something is to say you consider it important or worthy. Sometimes we value something by saying how much we would buy or sell it for, but this is way too narrow. After all, we might value a kiss from our grandmother but we wouldn't pay for it; more generally, a kiss that we pay for has a very different value than one we get for free. And we can value all sorts of things. We can value material things like cars, ephemeral things like moments of joy or pleasure, and symbolic things like an honor or award we receive. We can also place value on the things that we do. We value doing exercise or watching a film or having dinner or meditating. Values are necessary to make decisions. To make a decision, we must either figure out what our values are or make them up in order to decide.

Consider a simple decision. Should we shift our gaze to the right or left? We (or at least our brains) shift our gaze a few times per second when we're awake (and sometimes when we're asleep). The decision to look right might not be based on what we want to see; it could be based on nothing more than fatigue of the muscles involved in moving to the left. In that case, we are placing a value on the outcome of maximizing the amount of energy we have remaining. But the decision could also be dictated by an action value. Our eyes could be following a scripted pattern of saccades (eye movements) that decrees it is time to move to the right. In this case, the value determining the choice concerns the action itself. Even some simple decisions are made because of the values we place on particular actions.

When are values sacred? They must be about actions and not outcomes. But that is not sufficient. Some actions are not sacred. When I put on a shirt, my tendency is to insert my right arm first. This is an action, but I don't do it because of any sacred value; it is mere habit. I do it that way because my mommy taught me to do it that way and that's how I have done it ever since. On the rare occasion that I go to a fancy restaurant, I wear one of my two nice shirts. That's a value associated with an action too, not an outcome, but it is not a sacred value. I understand the fancy restaurant convention where I live to be that you should wear nice clothing, not my usual academic-leisure attire. Habits and conventions determine how we act, yet they are not necessarily sacred values, though sometimes they become sacred over generations in the way that many religions prescribe sacred head coverings (presumably many religious actions have their origin in simple convention). Notice that wearing a fig leaf to a fancy restaurant is different. Not showing your private parts is a sacred value in most corners of societies that have fancy restaurants, so wearing a fig leaf does violate most people's sacred values.

So how do we distinguish forces like habit and convention from sacred values? Sacred values are unique in being absolute; as psychologists Jonathan Baron and Mark Spranca put it, they are "rules that apply to certain behavior 'whatever the consequences.'"[5] In other words, sacred values don't trade off with outcomes of their associated actions. If you hold saving forests as a sacred value, then there is no amount of money I can pay you to destroy a forest. If you hold the right to own handguns to be a sacred value, then you will not be dissuaded no matter how many people die from guns (or bullets). This is what actor Charlton Heston was saying when he talked about prying his guns from his "cold, dead hands" in his presidential address to the National Rifle Association in 2000.[6] And in fact, when people are asked to make such trade-offs, they often refuse to even answer the question.[7]

Are Sacred Values Necessarily Moral?

You would be forgiven if, at this point in the book, you thought the topic of sacred values was all about how we make moral decisions. And much of it is. Yet the ideas are broader than that. They concern decision making in general, not just moral decision making. I am not even convinced that moral decision making is a well-defined field. Certainly moral considerations enter into many decisions. I would hope that they enter into a decision like, "Will I travel on an airliner if I'm testing positive for a contagious disease?" But a decision like that is not purely moral. There are also nonmoral facts about disease transmission, disease severity, airplane design, time of year, alternatives to air travel, and more that should be taken into account.

Are there any purely moral decisions? A good candidate might be abortion since at first glance, it seems like a contest between two moral imperatives: do not kill and a person's right to choose. But even that is just a shallow gloss. Abortion decisions involve lots of other factors: Are there health considerations? What stage is the pregnancy? Is the fetus likely to survive? Who else will be affected by the decision? Are there financial considerations? What does the future hold for the parent and child?

Maybe a purer moral decision would be a religious one. Should I pray according to the dictates of my religion? Yet even that isn't purely moral. After all, it depends on my beliefs. If my religious faith isn't strong enough, the decision may not be moral at all. It may just be a question of habit and tradition, or conforming to other people's expectations to avoid an argument.

The critical question for this book is whether our discussion of sacred values and consequences applies to decisions that have nothing to do with morality. Clearly consequentialist considerations apply broadly. Choosing between chocolate and vanilla has no obvious moral implications, but consequences matter in the form of gustatory experience.

Do sacred values apply outside the moral domain? They do in at least one sense. They shape our habits, and habit is perhaps the primary determinant of our choices. Many people grow up in a tradition that forbids eating certain kinds of food (like pork). Even those who no longer abide by the tradition's values may still avoid the forbidden food merely because it seems distasteful as it has never been part of the person's menu.

Beyond mere habit, sacred values can emerge from our identities rather than our morals. Some people may see themselves as a teacher, healer, or artist. They pursue their career because it represents the kind of person they see themselves as ("I was born to teach/heal/create"), not because it is morally required. Such an action is a sacred value in the sense that it has the normative properties of sacred values. It concerns an action (pursuing a particular kind of job) without regard for material trade-offs (regardless of how much the job pays).

Some actions seem like they are driven by sacred values but perhaps they are not. Perhaps you are the type of person who would go regularly to your parents' graves even though you are an atheist. You are not doing it to go to heaven and there is no real moral compulsion. Perhaps you are doing it to realize your identity (you are the type of person who shows respect for your parents). Or perhaps you are doing it for consequentialist reasons. Maybe it gives you a deep sense of fulfillment. This is a tough case and remains enigmatic, at least in my mind.

Whether or not all sacred values have their source in morality, people treat them as signaling an individual's morality.[8] Behavioral scientists Tamar Kreps and Benoît Monin took statements like the following quotes from US president George H. W. Bush:

> We should establish a legal and orderly path for foreign workers to enter our country to work on a temporary basis. As a result, they won't have to try to sneak in.
>
> We should establish a legal and orderly path for foreign workers to enter our country to work on a temporary basis. . . . We need to uphold the great tradition of the melting pot that welcomes and assimilates new arrivals.

Kreps and Monin asked people to read each statement and then judge to what extent the issue is a moral one for the speaker. Sentences like the first one were judged as indicating less morality than sentences like the second. How do the sentences differ? The first is an example of a consequentialist statement. It describes the consequences of having a good immigration policy (foreigners won't sneak in). The second is an example of a statement about sacred values. It justifies the policy by appealing to a "great tradition." The authors argue that consequentialist statements show less appreciation for what is sacred and more self-interest. They are also less abstract.

Whatever the reason, mentioning a sacred value gives listeners a stronger sense that the speaker is guided by moral concerns than by mentioning a consequence. So even if sacred values are not necessarily moral themselves, they signal morality. That's why we have second thoughts about hanging around with someone who wears nothing but a fig leaf. Even if we are not put off by what we might see, we are concerned about that person's morality.

The Critical Distinction That Everyone Loves to Forget

When studying decision making, there are two sorts of questions to ask:

1. The normative question, What's the best decision or way of making a decision?
2. The descriptive question, How do people make decisions?

It's hard to think of a simpler or more obvious distinction. The normative advice "buy low and sell high" is clearly correct, assuming your goal is to make money. But even superstar investors fail to abide by the maxim on occasion, and some of us simpletons fail to abide by it on a regular basis. What we do and what we should do are clearly different.

Yet I frequently marvel at people's failure to appreciate the distinction, even expert theorists who should know better. I hear people say something to the effect of, "I don't worry about that; there's no point," as in, "I'm not bothered by what people say about me. Being bothered doesn't do any good." It may be a true normative claim that it doesn't do any good, but wouldn't most of us love to be able to ignore what people say about us? That ability does not come from the fact that being bothered does no good; it comes from some reserve of self esteem. This example does not necessarily involve sacred values, but it is a perfect conflation of normative and descriptive theory, explaining a descriptive fact by appealing to a normative theory.

This is an important issue because it addresses the question of human rationality. One useful definition of rationality is the convergence of normative and descriptive theory. A judgment or decision is rational if and only if what people say or do is consistent with what an acceptable normative theory says they should say or do. On this definition, it is rational to buy lottery tickets if and only if you are more likely to end up with more happiness or utility if you do buy the tickets than if you don't. There are lots of other senses of rationality, but this is the notion that I'll be using in this book. Let's consider sacred values first, briefly, from a normative point of view and then at greater length from a descriptive one.

Properties of Sacred Values: Normatively Speaking

The discussion so far has attributed two normative properties to sacred values: They apply to actions, not to outcomes, and they are absolute. These properties are closely related. If a value is absolute and therefore does not trade off with consequences, then it cannot be a value concerning an outcome (because consequences don't matter). Thus it must apply to the other component of decision making, actions. So the absoluteness of sacred values implies that they have to be about actions and not outcomes.

The absoluteness of sacred values can get us into trouble. Sometimes we are faced with competing sacred values (what I called tragic trade-offs in chapter 1). For instance, I might value not killing animals, so I plant a field of vegetables in order to enable me to avoid eating meat. But in the process of clearing the field, I am likely to end up killing dozens of rodents and other small life-forms that live in that field as well as possibly killing some of the birds that would have eaten the animals I've killed. My desire to avoid killing animals has caused me to kill animals. Or consider buying an electric car to reduce your carbon footprint (a sacred value). By doing so, you create more demand for the precious metals like lithium that are required to manufacture batteries. Extracting precious metals is also damaging to the environment and has led to ecological disasters and strife in South America and elsewhere. In cases like this, we must engage in trade-offs, not between sacred values and consequences, but between sacred values themselves.

Properties of Sacred Values: Descriptively Speaking

Descriptive questions about sacred values concern their psychology and how people treat them. Are people willing to call something a "sacred value" (or equivalent term) yet not treat it as absolute? Does thinking in terms of sacred values have other effects on what we believe, what we think, or how we feel?

Outrage. One reason that sacred values have an outsize effect is that violating them can elicit outrage. I value free speech and am outraged when I see speakers shouted down, even speakers I disagree with. Most everyone values life, and we are all outraged when we hear about a powerful government assaulting unarmed civilians. Observing a violation of a sacred value evokes strong negative affect, especially anger or disgust, the precursors to outrage.[9] And outrage is the first step in punishing others for their wrongdoing.[10] In other words, human society has evolved an effective mechanism for preventing violations of sacred values: Third parties who observe such violations become angered or disgusted and then outraged, and this leads them to want to punish the violator. Notice that this happens without much thought. Some thought is required just to comprehend what's going on. Beyond that, the mechanism operates through affective channels more than cognitive ones.[11]

As usual, there are exceptions. Violations of sacred values can occur simply by thinking wrong thoughts. This is common religious doctrine, exemplified by Matthew 5:28: "But I say to you that everyone who looks at a woman with lustful intent has already committed adultery with her in his heart." Jimmy Carter knew his Bible and admitted having engaged in "lust in his heart" in a *Playboy* interview during his US presidential campaign in 1976. It almost torpedoed it.[12] Carter was raked through the coals for his impure thoughts. Author Salman Rushdie was sentenced to death by the Islamic regime in Iran simply for inviting readers of his book *The Satanic Verses* to imagine the Prophet Muḥammad keeping the company of prostitutes.[13] In modern Western society, repetitive impure thoughts are a classic symptom of obsessive-compulsive behavior that can lead to torment, anguish, and shame.

Eliciting outrage matters. Today—the mid-2020s in the United States—it is what makes the country run. In his book *Why We're Polarized*, Ezra Klein makes a strong argument that the road to political success in the United States has little to do with constructing effective policies, demonstrating political prowess, or even kissing babies.[14] It is the ability to elicit outrage. To get political support, he says, don't solicit support; rather, make your voters hate the other side. Anger is what wins elections. If he's right, then a recommended strategy would be to show how the other side is violating your sacred values. A politician might accuse the opposition of, say, allowing illegal immigrants into the country. Or if speaking to a different constituency with different sacred values, a politician might accuse the other side of abusing long-standing power relationships. Both accusations are attempts to elicit outrage in order to foster hate.

Action. Sacred values motivate people to act. Hyoseok Kim and I asked people about a policy that had been studied by Phil Fernbach and Lauren Min, allowing lion hunting for a fee that, in part, supports lion conservation. We had people frame the issue either in terms of sacred values or consequences. We could not simply ask them to think about the issues in one of the two ways because the distinction is subtle and most people do not understand it at a conscious level. So we had to be tricky. We asked people to come up with a reason for their attitude toward the policy by constructing a sentence that used one of six words. Half the subjects generated a sentence that used "value," "freedom," "respect," "true, "harm," and "moral." These words are closely tied to sacred values so we expected people to generate a sacred value that served to frame the issue. The other half of the subjects were asked to construct a sentence and were given the words "consequence, "cause," "effect," "cost," "benefit," and "outcome." These words are closely related to consequentialist reasoning and so we expected a consequentialist sentence to serve as a frame. The manipulation worked in the sense that most people—though not everyone—generated the kind of reason we expected. Then we asked everyone whether they wanted to donate to a charity that supported their position on the issue, and if so, how much they wanted to donate. More people who generated a sacred value reason were willing to donate, and donate more, than those who generated a consequentialist reason. The sacred values frame was more effective at spurring a willingness to act. We have tried this procedure with several issues. The effect occurred in most cases.

Fernbach and Min had found that those who report assuming a sacred values frame, rather than a consequentialist one, are more likely to consider an issue relatively simple and well understood. This provides an account of why people are more willing to act under a sacred values frame; action is more justified when the basis for action is understood. Sacred values frames induce people to take more extreme positions on the issue too, and to see the issue as less tractable and less likely that conflicting views on it will be resolved. Fernbach and Min also ran studies that trained people on the sacred values versus consequences distinction and then asked participants to elaborate on one or the other. Those participants who elaborated on sacred values judged resolution to be less tractable than those who elaborated on consequences. Moreover, when they presented participants with experimenter-generated arguments, participants who read arguments based on sacred values judged resolution to be less tractable, were more certain about their position, and became more extreme in their position, compared to participants who read consequence-based arguments. These studies suggest that thinking about issues in terms of sacred values, as opposed to consequences, causes people to think the issues are simpler, that they understand them better, and that they will not compromise on them. After all, consequences are complex and messy. Sacred values are simple. Violating a simple truth is more likely to outrage someone than saying something complex and nuanced.

Cleansing. Phil Tetlock and his colleagues point out a property of sacred values that reflects the feelings of disgust that emerge from sacred value violations—the desire to "cleanse." They asked people how willing they were to join a political group supporting a sacred action. Subjects were asked either before or after having engaged in a set of tragic trade-offs (like whether to save one child or two adults). Tragic trade-offs require one to violate a sacred value in order to maintain the other. Subjects were more willing to join the political group after having engaged in the trade-off than before they had engaged. Their greater willingness to take an action consistent with their sacred values after engaging could arise from a desire to cleanse themselves after violating a sacred value, not unlike the practice in some religions of suffering or self-flagellation to help purge one's sins.

People are generally willing to trade sacred values off with each other. Jonathan Baron and Sarah Leshner pit policies representing conflicting sacred values against each other. In one case, subjects were asked to imagine they had to negotiate cuts to two policies reflecting sacred values. The policies might concern a plan for medical aid to the poor for lifesaving cancer surgery and a plan preventing the loss of endangered plant and animal species. Two-thirds of the time, their subjects had no trouble making bigger cuts in one policy than the other; some values were deemed more important than others. Perhaps it seems like a logical violation that two values that are both deemed absolute can be traded off with each other. But it doesn't violate intuition. Most people who do not eat pork would encourage their family to if it was the only way to prevent starvation. And those of us who care a lot about the environment would drive a gas-guzzler if it was the only vehicle available to get a person in need to a hospital. Everyone might order the significance of sacred values in their own way, but some tragic trade-offs are inevitable.

Taboo trade-offs. In contrast, people are rarely willing to engage in trade-offs between sacred values and material benefits. In another Baron and Leshner study, few people could think of benefits that would be so great that they would be willing to violate their own sacred values like not buying clothes from a manufacturer that uses young children at sewing machines.[15] Sometimes they could, though. For instance, they might have considered the possibility that the labor was providing sustenance to the child's entire family and come to the conclusion that saving a family justified the child's labor. When they did have these insights, they almost always decided the values weren't sacred after all. What is clear is that there are some people who will not trade material goods for sacred values.[16]

Nothing is ever simple, however. In other experiments, Baron and Leshner asked about a new type of genetically engineered wheat that had both potential to do harm (e.g., damaging the environment) and create benefits (e.g., reducing malnutrition). The experimenters varied the likelihood of these harms and benefits. Subjects might be told that there is a certain probability of harm (e.g., only 1 in 10,000,000 versus 1 in 100,000), but the benefit is guaranteed. When asked whether the wheat should be approved for use, evaluations of this supposedly absolute sacred value turned out to be sensitive to the probabilities of harm and benefit. Subjects were most likely to approve the wheat when the probability of harm was lowest and possible benefit was highest. Subjects thought their judgment was based on sacred values but in fact they were influenced by consequences.

Sacred values are not supposed to be sensitive to consequences, so the probability as well as the amount of harm and benefit should not matter. But it can. Baron and Leshner's interpretation is that "[sacred values] are strong opinions, weakly held." But how can that be? Isn't a strong opinion by definition the kind that one holds onto come hell or high water? Apparently not. Here's one possibility: Sacred values are strongly held, but only when they serve to frame the problem and when the cognitive system uses them as a framework for understanding. Other frames are possible too. When people are thinking in terms of consequences (frames that are sensitive to benefits and harms), then sacred values are out of the picture. Yet when we are thinking with a sacred values frame, consequences be damned.

Deceptive values. Sacred values reduce the need to engage in taboo tradeoffs, but not all values are as they seem. Some of our values only appear to be sacred but are imposters. We might assert that we always tell the truth, yet in the face of the possibility of serious consequences, we bend it (e.g., to save a friend's life or reputation). Sometimes we take on sacred values to signal our virtues. I am a wonderful person because I make generous charitable donations, though I might only make them when I'm rewarded with public recognition. In both of these cases, the sacred values may be more apparent than real. People are willing to trade them off for consequences. Actually, assuming a sacred value to signal a virtue may not be wholly perfidious. Once you put a sacred value in your cap for all to see, then social forces may cause you to abide by it in order to satisfy a different sacred value, avoiding hypocrisy.

Omission bias. Which is worse, to be walking down the road with a sandwich and not give it to a hungry beggar, or to actively prevent a do-gooder from giving a sandwich to a hungry beggar? Is doing nothing (omission of an action) better or worse than taking an action that has the same consequences (commission)? In many circumstances, people judge commission worse than omission. You probably agree in this case. We all fail to give hungry beggars sandwiches, but it verges on evil to get in the way of somebody else feeding the hungry. This tendency to judge that commission is worse than omission is called omission bias. Ilana Ritov and Jonathan Baron argue that omission bias is particularly strong when a sacred value is at play. For example, they consider a case of an epidemic that would cause 10 out of 10,000 children to die, and ask whether a vaccine should be administered that would prevent these deaths but would itself cause 5 out of 10,000 children to die. They found that many of their participants opposed the vaccine—despite its better consequences. The researchers suggest that the reason for this is that violating a sacred value by commission is worse than violating it by omission.

They also tested scenarios like the following:

> A convoy of food trucks is on its way to a refugee camp during a famine in Africa. (Airplanes cannot be used.) You find that a second camp has even more refugees. If you tell the convoy to go to the second camp instead of the first, you will save 1000 people from death, but 100 people in the first camp will die as a result.
>
> Would you send the convoy to the second camp? Y N
>
> What is the largest number of deaths in the first camp at which you would send the convoy to the second camp?[17]

Those who thought preventing starvation was a sacred value gave a lower requirement for the number of deaths in the first camp than those who did not consider it a sacred value. In other words, those reasoning with a sacred value have a stronger preference for omission over commission than those who are not. This may be because sacred values are generally prohibitions against actions.[18] We are prohibited from actions that contribute to the death of children. As a result, taking an action that results in their death is worse than not taking an action that has the same result. In the latter case, we are not responsible because the cause of death is starvation. Some sacred values are not prohibitions against action. "Pray facing Jerusalem" is a sacred value that promotes a particular action. Perhaps the omission bias would not be observed with such values.

Insensitivity to quantity. A property of sacred values that derives from their absoluteness is that any violation is wrong, regardless of the extent of the violation. People who hold that destroying forests violates a sacred value indicate that it is just as bad to destroy a large forest as a small one.[19] Similarly, most people believe that employing a ten-year-old in a garment factory for sixty hours/week violates a sacred value. They also think that employing 125 ten-year-olds is worse than employing 1, but not much worse (a lot less than 125 times as bad).[20] Sacred values are, in this sense, treated as pretty much absolute.

One source of evidence about quantity insensitivity comes from studies like Ritov and Baron's that ask whether people would engage in an action that would violate a sacred value (like rerouting a convoy of food trucks, leading to 100 deaths) even if the action had more positive consequences than inaction (because the action would save 1,000 people). How many people would have to be saved to justify rerouting the convoy? One set of studies shows that people with a sacred value (like preventing starvation) are insensitive to the quantity of deaths and thus require that fewer people would die in the first camp than those without such a sacred value.[21] But it turns out that the result depends on how you ask the question. Psychologists Dan Bartels and Doug Medin found that when they asked a series of questions akin to, "Would you reroute the convoy if 100 people would die in the first camp?" "Would you reroute it if 200 would die?," and so on, then people with sacred values said yes to a higher number, not a lower one.[22] Bartels and Medin argue that whether or not one is sensitive to quantity depends on what one is attending to. I agree. The original procedure frames the issue with attention to sacred values; it asks people what outcome would justify violating their sacred value. In contrast, the Bartels and Medin procedure frames the issue with attention to consequences (what are the outcomes of acting and not acting?). How we think about issues is not fixed. Sacred values and consequences provide different frames, and we can induce people to see an issue either way.

The sacred veil of ignorance. Sacred values are critical for a healthy society, but they can bias our judgment. When predicting the future or explaining the past, we rely on judgments of probability, the likelihood that an event occurred. These judgments are critical for decisions we make. An insurance broker needs to know the probability of burglary and fire before determining how much to charge for home insurance. It turns out that it matters where those probabilities come from. Tetlock and colleagues told people that an insurance underwriter had determined that some towns had a 10 percent probability of a break-in or fire and other towns had less than a 1 percent chance.[23] Most people, including liberals, showed some support for differential pricing—charging towns a greater premium if they had more risk. Another group of subjects was told that the towns also differed in their racial composition; the towns with higher risk had a greater proportion of African Americans. Now liberals were much less likely to support differential pricing (moderates and conservatives were not affected much). For liberals, not discriminating based on skin color is a sacred value and differential pricing violates that value, so they rejected it. But the fact that the towns differed in their composition does not change the probability of a break-in or fire and, as such, the actual risk faced by the company in each town. Thus making the sacred value salient had the effect of causing liberals to reduce their attention to risk. Their judgments were instead influenced by the sacred value itself.

Outcomes. Sacred values might, ironically, even affect how we perceive outcomes, at least legal outcomes. Professor of psychology Linda Skitka at the University of Illinois at Chicago and colleagues have argued that whether or not people perceive a punishment in the courtroom to be just is not determined by the fairness of the procedures used to try the case.[24] Rather, people only consider a verdict just if the guilty are penalized and the innocent acquitted. People's reactions to the outcome of the O. J. Simpson murder trial had little to do with their assessment of how fair the trial was; the reactions reflected whether or not they thought O. J. was guilty.

We hold sacred that only the guilty should be punished. We tend not to think carefully about how outcomes were decided despite the importance of having a just procedure for determining guilt and punishment.

Social cohesion. Sacred values have always been central to fusing individuals into communities. They bring people together by providing both a common ground and common enemies, those who defy the common ground. Sacred values worked this way in ancient Greece; Socrates was condemned to death for not sharing the values of ancient Athens (he was a critic of democracy and rejected the city's gods). And they work this way everywhere, including in the modern-day United States. For this reason, if you're in the position of having to confront or negotiate with someone from a different community (if, for instance, you are a US congressperson and want to get something done in a bipartisan fashion), then it turns out that it is critical to acknowledge and show that you understand the other side's sacred values. You don't have to agree with them, but you have to recognize them publicly.[25] Social psychologist Jonathan Haidt developed a theory of US sacred values that he called moral foundations. He proposed that liberals in the United States value caring, fairness, and liberty. Conservatives, he contended, also maintain those values and, in addition, value loyalty, authority, and sanctity.[26] In more recent work, fairness has been broken down into equality (that everyone should be treated equally as well as have equal say and opportunity) and proportionality (that social rewards and punishments for everyone should be proportional to their costs, and the person's contributions, effort, merit, or guilt).[27] On the more recent taxonomy, conservatives tend to score higher on loyalty, authority, purity, and proportionality, and liberals score higher on care and equality.

Sacred values are the wellspring of our social identities, and not just because we identify with others who share our values. This was pointed out by French existentialist Jean-Paul Sartre in a discussion of the unexpected benefits of the Nazi incursion into France in World War II: "The choice that each one made of his life and of himself was authentic; for the more it was made in the presence of death, the more it could always be explained in the formula of 'Rather Death Than . . .' And I do not speak here of our elite who were the true Resisters, but of all Frenchmen who, at every hour of the day or night, during four years, have said NO. . . . Each of them set himself freely, irremediably, against the oppressors. And in his freedom in choosing himself, he chose the freedom of all."[28] Sartre claims that by risking death at the hands of the Nazis to assert their freedom, the French were expressing a sacred value that pertained not only to their individual being but to all French people, and perhaps even beyond France. Fighting for a sacred value establishes a deep sense of community with all who share that sacred value.

The cohesive role of sacred values is also demonstrated by its exceptions. Psychopaths are not social butterflies, and indeed have a lot of trouble making friends and fitting in socially.[29] Presumably this has a lot to do with their core diagnostic characteristic: They lack empathy for others, especially affective empathy, the ability to experience others' emotions and pain. When asked to make judgments about moral dilemmas that involve harming others, psychopathic individuals do not find these conundrums to be as morally repugnant as others do.[30] Indeed, there are a number of studies showing that psychopaths' choices in such problems tend to be guided by concern about consequences, not by sacred values. The fact that psychopathic individuals can be so driven by outcomes may be one reason that they are often treated as pariahs.

People actually like those who reason with sacred values more than they do consequentialists. They trust them more and value them more as social partners.[31] Consequentialist reasoning comes across as cold and calculating. In the 1970s, the Ford Motor Company ran into trouble when one of its vehicles, the Pinto, caught fire several times as a result of rear-end collisions. The design of the car put the gas tank in a vulnerable spot that could leak fuel and cause a fire if hit from behind. The public reaction against Ford was titanic, spurned to a great extent by an article that appeared in *Mother Jones* magazine accusing Ford of trading lives for profit. The article probably got the details wrong, but it is true that Ford reported a cost-benefit analysis in a government filing that compared the cost of recall and modifications to the societal cost of the deaths and injuries that would result if no changes were made. Ford determined the cost to be lower if it simply did not fix the engine.[32] It was this consequentialist reasoning that produced the public outrage. A car company that put a dollar value on human lives to determine how much risk to take could not be trusted (even though health insurance companies do this every day). The problem with the sacred value that generates this reaction is that if human lives are deemed worth sparing any expense for, then no car would be permitted, nor would any contact sport, crossing the street, or indeed anything that poses any risk to human life. Living a life worth living requires trade-offs.

In contrast to consequentialist reasoning, sacred values reasoning gives the impression that the reasoner has a conscience. That is why politicians are so happy to assert their sacred values. Both the Republican and Democratic Party platforms start with a list of sacred values ("liberty, economic prosperity, preserving American values and traditions, and restoring the American dream" for Republicans, and a longer, more specific list for Democrats).[33] Politicians are particularly happy to spout sacred values when they are ignorant of the facts. US vice presidential wannabe Sarah Palin was ignorant enough to claim that 9/11 was carried out by Saddam Hussein. But her lack of knowledge did not reduce her faith in her values. She also claimed her nomination was "God's plan."[34] Asserting our sacred values is the fast road to showing off our virtue. How can someone who tells us how much they value equality or freedom not be a good person?

The answer is that it is easy to value equality or freedom. It's even easy to value both. What is hard is making decisions in a complex world full of uncertainty, ambiguity, and competing values. I am less impressed by vows of equality or freedom than I am by the ability to, say, develop a top-notch company or program while simultaneously showing respect for all concerned, allowing a diversity of perspectives, and making everyone feel heard. I believe that we can and should think consequentially, and can do so without giving up on all (or possibly any) of our sacred values. It is a sad truth that people avoid consequentialist reasoning in order to get along with others, and that society sometimes requires an appeal to sacred values and abhors cost-benefit analysis. But as we will see, some form of consequentialist analysis is the only way to solve society's most pressing problems.

The Locus of Sacred Values

Where do sacred values live? Sometimes they are written down like the Ten Commandments. But more often they are cultural artifacts; they are represented and practiced by a culture in myriad ways, not relying on any concrete form. What form do your values about climate change take? What is their source, and what provides them sustenance? I posit two answers to these questions.

First, sacred values are embodied in cultural narratives. Every culture has a narrative that describes its origins and history, prescribes its values, and offers anecdotes that exemplify its values and conventions. This narrative serves as the backdrop for cultural events and objects. In Native American cultures, for instance, it is common to view creation as a living process, with Mother Earth, the waters, and the sun as living beings, and all creatures as sharing some form of kinship.[35] Out of these conceptions flow stories of creation and seasonal transformation, values about how to treat nature and other people, and beliefs about what causes change. Narratives are not static but instead constantly evolving as events transpire, and technology and the environment change. The United States is constantly debating its cultural narrative, and the result is a narrative today that is a far cry from the one I grew up with—that the Pilgrims ate turkey with the Indians, George Washington never told a lie, and mother is in the kitchen and father in the office. At the heart of cultural narratives are sacred values, for they serve as the motivation for key players' behavior (George Washington valued the truth just like a good American should) and frequently to motivate the growth over time of the community itself. Russia's narrative exclaims its value as an exceptional nation by portraying itself as an innocent victim of foreign attack, before heroic triumph avoids total defeat.[36]

Besides being housed in narratives, sacred values are embodied in cultural practices. The world is inordinately complex, and appropriate action always depends on subtle aspects of the situation. Whether or not to take a sip of the delicious wine in your glass depends on your religion, whether or not the right blessing has been uttered or a toast has been made, the time of day, whether or not you're at the king's table, and much more. Individual humans do not have sufficient knowledge or processing capacity to make all of these judgments and decisions without the help of others.[37] Decision making—and thought more generally—relies on others.

Much of the support we get from others comes from our cultural institutions. Hospitals help us with health issues, government agencies and courts help us with legal issues (when they're not hindering us), and museums help us learn about and give us the opportunity to appreciate art. Humanity could not navigate the social and moral world without institutions to guide us.[38] And institutions embody our sacred values. Hospitals realize our goal of saving lives, the legal system is designed in part in an effort to codify sacred moral values, and museums help to represent and transmit sacred values. It is these institutions that create and shape our cultural practices, and thus shape and enforce our sacred values.

The point here is that sacred values are not merely ideas that sit in individual brains. They are not mere psychological entities. We need not even think of them as belonging to individuals. Sacred values drive decision making by virtue of being shared by a community in its narratives and cultural practices. The freedom to vote isn't just an idea, it's a practice enforced—more or less—by electoral commissions and it gets its justification from stories about revolutionary acts that forced a small-minded elite to broaden the franchise, like the suffragette and civil rights movements in the United States. Sacred values support social cohesion because they are primarily social entities.

Essential Sacred Values

This is one reason that certain sacred values are essential. They serve as a signal that one is moral, and being moral is a requirement of most communities because it is a necessary condition for trust. It is hard to cooperate with someone you do not trust. Someone without any moral strictures is free to lie and betray, and therefore cannot be trusted. Even communities of gangsters have to trust one another, and so signals of morality are crucial. Successful gangsters have strong sacred values involving loyalty.

But not all sacred values relate in obvious ways to trust and cooperation.[39] Sacred values like temperance, sobriety, asceticism, and modesty are about purity, avoiding too much (or any) alcohol, other drugs, any form of indulgence, or sex. Piety, ensuring sufficient reverence, is another example. All of these sacred values seem to pertain to actions that primarily affect the individual acting, not anyone else. Why should communities care about them? Why are such values central elements of so many cultures? A number of theories have been offered. One is that avoiding drugs and sex is a good way to avoid pathogens.[40] Another idea is that having sacred values about such things helps people to choose which side to take in a dispute.[41] A third holds that puritanism of this sort evolved to enforce self-regulation.[42] The idea is that it is good for us to refrain from indulging in vices that are pleasurable. Doing so is one way to improve our self-control. This is important because self-control is the key to cooperating with others. After all, we cannot help others if we are too drunk or stoned, and we will only foster conflict and resentment if we are too promiscuous. Whatever the reason for the prevalence of puritanical sacred values, all of these views argue that they serve a social function. They minimize conflict while maximizing health and cooperation within our communities.

Sophisticated human society cannot function without sacred values. We can see this from a sacred values perspective: It is only right that I and the people around me must conform to some minimally acceptable standards of behavior. We can also see it from a consequentialist perspective: We can only trust others and expect the best from them if they hold at least some values to be sacred. Sacred values provide good reasons to open hospitals and feed the needy. The fact that sacred values are essential for a functioning community turns out to be the first horn of a dilemma. We will see the second horn in the next chapter as I review how much trouble this dependence on sacred values gets us into.

3 Why Sacred Values Are Dangerous

Paul Jennings Hill had a sacred value. He believed that abortion was murder. "I definitely felt that the Lord wanted me to shoot the abortionists," he said after killing physician John Britton and Britton's bodyguard, retired US Air Force lieutenant colonel James Barrett, and wounding Barrett's wife outside an abortion clinic in June 1994.[1] Hill's last words before being executed by lethal injection were, "If you believe abortion is a lethal force, you should oppose the force and do what you have to do to stop it. May God help you to protect the unborn as you would want to be protected."[2] Hill's sacred values led him down a rabbit hole that almost everybody considers misguided and wrong. Antiabortion crusaders themselves rarely condone murder.

Sacred values may signify that one has a conscience, but they also have a dark side. They can lead to extremism and terror. They can be the foundation for an intransigence that is a root cause of many of society's deepest and most pressing problems.

Heresy

These problems go beyond the risk that sacred values will take hold of an individual and make them a zealot. Individual acts of malevolent extremism like Hill's are horrifying, but mercifully rare. That's not the direction that sacred values normally press. More commonly, sacred values operate by binding members of a community and encouraging group action. In the last chapter, we saw that communities are often defined by shared sacred values. Religious communities, for example, are replete with sacred values of various kinds, including moral proscriptions, ritual acts, and belief in deities. Adherence to such values is often necessary and sometimes sufficient to be a member in good standing of a religious community. For centuries, the vast majority of Europeans were under the sway of the pope, who could (and would) excommunicate antagonists who advocated beliefs inconsistent with church doctrine. But social pressure to conform to community sacred values is not limited to Christian Europe. Even the rare case of extreme punishment can be observed in far-flung corners of the world. The Pakistani Taliban shot Malala Yousafzai in the head because it did not approve of her campaign in support of girls' education. Milder forms of punishment for violating a community's sacred values are common, if not universal. Not only must sacred values not be violated to remain a citizen in good standing in your community, but they should not even be contested.

We see it today in the polarized United States. On the right, people have been accused of being Republican in Name Only (RINO) since the early 1990s—an epithet intended to suggest the accused do not share the party's ideology, including its key values. The term was stock-in-trade for Donald Trump, although his usage frequently had nothing to do with any political value other than reverence for Trump himself. On the left, "cancel culture" is all about condemning people and forcing them out of the community for not being sufficiently progressive in their sacred values. It is hard to gauge the actual extent of cancel culture because it has become one of the totems that the Right has used to characterize and condemn the Left. Thus its main effect may have been to polarize rather than change minds.

Sticking with your community is generally an adaptive strategy. It prevents you from becoming ostracized and affords you the safety of numbers. It is safer not only in that having people around you protects you from attack by other animals (especially other humans) but also in that it allows you to take advantage of other people's labor. Cooperative groups provide far better for their members than individuals trying to make out on their own (Tarzan excepted).

The Power of Simplicity

Sticking with your community drives you away from other communities because communities often feel the need to distinguish themselves. A community can distinguish itself by achieving scientific breakthroughs, winning gold at the Olympics, or providing high-quality support services. But it can also distinguish itself by putting other communities down, discriminating against members of other groups, or engaging in battle. Battles can be psychological or physical, economic or violent. Psychological battles include the efforts to dehumanize enemies that history is studded with, from the ancient Romans lording it over the ancient Gauls (as parodied in the Asterix comic books) to the European desecration and destruction of Indigenous culture throughout the Americas, to the Nazi caricatures of Jews, Roma, and Slavs. Each of these examples (and so many more) involves dehumanization along with economic inequity and physical brutality, brought on not only by competing interests but also by humanity's habit of dividing into competing groups based on clashing narratives.

Sacred values serve a community's narrative the way a constitution serves a body politic or a holy book serves a religious group. Knights in the Middle Ages struggled for honor, Nazis fought for their imagined race, the French for "liberté, égalité, fraternité," and (most) Americans to spread democracy. Communities like to say that they are fighting for an idea—for a sacred value. That can make sacred values a central pillar of much division and hate. As political scientist Morgan Marietta says about the history of the United States, "The political rhetoric that citizens encounter is often unconflicted, extreme, and strident, taking positions that ignore compromise or negotiation, upholding the inviolability of a favored set of values while dismissing others."[3] This isn't true just in the US; the rhetoric of sacred values governs history the world over.

Sacred Simplicity

At the time I'm writing, the United States has been riven by identity politics for the better part of a decade. On the right, books are banned and academic ideas bound by the label "critical race theory" are outlawed in public schools in many conservative states. The legislation known as the Individual Freedom Act took effect on July 1, 2022, in Florida and is known on the streets as the "Stop WOKE Act." It purports to protect the open exchange of ideas, but is clearly intended to prevent teaching a progressive-left perspective whose basic premise is that the US is deeply racist. On the left, there is a movement to "decolonize" thought, including changing Western educational systems from the ground up.[4] The imagined enemy is the hegemony of Western systems of knowledge perpetuated through the domination of white males. Through its colonial history, Western thought apparently won the battle of ideas not through any inherent accuracy and usefulness but rather through social and political power. Western scientific practice is just one way of knowing among many, and it should yield its position of dominance to ways of knowing developed by other cultures. The fact that Western science has put people on the moon does not make it special; after all, Eastern meditative practice has the ability to make people at one with the universe.

What does this have to do with sacred values? On the one hand, it is not about sacred values. Some of it is about the distribution of power and resources, like money, jobs, and education. And under this framework, it makes sense that white supremacists tend to be white and African Americans vote disproportionately for progressive Democrats. But what is striking about the culture wars is how much lobbying people do against their own interests. Whites suffering from extreme poverty depend on government programs to stay afloat, but support politicians who cut those programs. And on the "Left Coast," white men are at the vanguard of the progressive movement, pushing hard for diversity, equity, and inclusion.

These people are not acting in their own direct material interest. Instead, they are called to action by the ideas endorsed by their communities. The Right is called to conserve or reestablish the true America (Make America Great Again). The Left is called to build a United States based on diversity and inclusion. Both of these calls are for action. As soon as they are seen in absolute terms, as among a citizen's highest callings, they have become sacred values.

What is it about sacred values that gives them such power? The key, I believe, is that they seem to eliminate the need to make trade-offs. Figuring out how to make trade-offs is the hardest part of making a decision. It requires figuring out exactly how much each aspect of a decision matters to you. When buying a house, how much distance from work would you be willing to trade for each foot of extra floor space? There is no recipe for making that trade-off, and when it matters, it requires a lot of soul-searching. But once you have a sacred value in hand, trade-offs with material consequences are no longer required. You only need to appeal to the action rule supplied by the sacred value. Those who don't drink don't have to figure out if they have had enough, and those who don't eat meat don't have to decide if that expensive cut of sirloin is worth it. Having a sacred value has the added benefit of simplifying decision making while providing a normative justification: "I'm not engaging in trade-offs. It's not because I'm lazy but because it is the wrong thing to do!" This makes decision making easier while making the decision maker feel more satisfied by doing the right thing, behaving consistently with their values.

So extremism is a kind of cop-out. Extremists rely on sacred values because that is a way of avoiding difficult trade-offs. This conclusion dovetails with studies of human motivation.[5] One prominent motivational theory is that extremism results from a single need coming to dominate all others, overriding other basic concerns. People always have multiple needs to attend to (adequate food, sleep, love, stimulation, etc.). Moderates strive to strike a balance among all of these needs. But an extreme dieter has put their need to lose weight above all else, an extreme infatuation puts a particular love interest above all else, and addictions can put a craving above other desires. It is similar for sacred values. A terrorist compelled by a sacred value has put their need to strike out above their own and everyone else's needs. Rather than engage with the struggle to balance a variety of concerns, extremists just focus on one.

Sacred values provide psychological comfort in other ways as well. Because they are absolute, they offer a kind of certainty. And because they simplify, they offer a way to reduce confusion. In *The Way Out*, author Peter Coleman makes the following observation: "When life gets particularly tense, unpredictable, or dangerous, we seek consistency and certainty even more desperately. So it follows that in more threatening times, like today, we find comfort and solace in moral certainty." Moreover, under threat, "we become more contemptuous of the other side."[6] This leads to his advice to deal with uncertainty and confusion: Don't contradict yourself, don't violate your own values, and don't challenge the beliefs of your community. Aligning yourself with your community by celebrating its sacred values will leave you feeling better about yourself because you will have more confidence, feel more knowledgeable, and feel more solidarity with the people around you. It seems like a win-win. Of course, it fails to appreciate that greater solidarity with your compatriots means less solidarity with out-group members, greater polarization, and—in the extreme—a society at war with itself.

These dynamics have, again, been described eloquently by Marietta: "Greater invocations of moral outrage engender a more strident form of politics."[7] When our reasoning and discourse is grounded in absolute notions of correct behavior, we become conjoined with our own ideas, identified with them, and therefore unable to compromise as well as deliberate and reach agreement with others. This increases political intensity and engagement.

Gun rights advocates appeal to a sacred value ground on their view in the Second Amendment of the US Constitution, "a well regulated Militia, being necessary to the security of a free State, the right of the people to keep and bear Arms, shall not be infringed." Grounding their argument in the Constitution allows them to support their position with a sacred value, making the issue seem simple and understandable, and their position fully justified. Given what the Second Amendment actually says, the simplicity of the justification is illusory. Many of the laws that they advocate, like concealed carry rights or the right to automatic weapons, do not follow in any obvious way from the Second Amendment but instead require complicated legal arguments.[8] Indeed, a rather direct reading of the amendment is that it only applies to militias and not to individuals. But gun control advocates have a fatal rhetorical disadvantage. They don't have an obvious and strong sacred value to appeal to, only a complex tangle of arguments and data.

The heart of the gun control argument rests on consequences. Guns are part of the chain of causation that led to over forty-five thousand deaths in 2020, including murders, suicides, accidents, killings by law enforcement, and those whose circumstances could not be determined.[9] What was the cause of those deaths? Was it the bullet, the gun, or the mental state of the person who pulled the trigger? To argue for gun control requires working through these issues and many others. And it's complicated by the data. Some of the data show that even though the number of guns in the United States has grown so that there are more guns than people, firearm homicides dropped about 40 percent between 1993 and 2018.[10] So the data do not provide unequivocal support for gun control measures. Studying the effect of such measures is not easy. The number of incidents is enormous by any measure of personal catastrophe, but nevertheless small from a statistical perspective, so that clear answers are not forthcoming. Moreover, the data have to be collected amid a variety of changing social norms and outbreaks of violence that vary by region, making it difficult to isolate the effect of any single type of intervention. Pitting the many-headed beast of facts about the consequences of antigun measures against the simple purity of a sacred value in favor of the freedom to carry is not a fair fight. The appeal to sacred values makes it seem simple enough for kindergarten.

Sacred or Moral?

Psychologist Linda Skitka has been at the center of research demonstrating that people who experience moral conviction about an issue are more politically engaged with the issue. She measures moral conviction using a couple of five-point scales, asking people questions akin to, "To what extent is your position a reflection of your core moral beliefs and convictions?" "To what extent is your position connected to your beliefs about fundamental right and wrong?" Not surprisingly, people consider attitudes that they rate highly to be universally and objectively true. People with such attitudes also tend to be intolerant of the other side and less willing to compromise than others.[11]

To illustrate the latter point, political scientist Timothy Ryan asked a group of subjects Skitka's moral conviction questions about Social Security in the US, presenting them with a Republican and Democratic position on the issue:

> As you may know, the Social Security program in the United States is projected to run out of funds in 2033 if changes are not made. One idea that has been proposed to address the problem is to raise taxes on people currently in the work force. Alternatively, some people have proposed cutting back on the benefits the government provides future retirees. How about you? Would you prefer to see taxes raised to preserve benefits at the current level, or would you prefer to cut benefits so taxes don't have to go up?[12]

Participants with more moralized attitudes were more resistant to compromise on the issue. In fact, they even wanted to punish politicians who were willing to compromise. These moral convictions have the trappings of sacred values. Ryan gave subjects the option of earning extra money if they allowed the researchers to donate to a political lobby group that opposed their moral conviction. Relatively few people with strong moral convictions accepted the offer. As we have seen among people who hold sacred values, they were unwilling to make material trade-offs—to accept money—if it violated their moral values.

Other studies have shown related effects on different kinds of issues. One focused on the value of technology.[13] People were presented with different messages containing arguments opposing crime surveillance technologies or hiring algorithms. The arguments either had moral content or focused on the financial benefits of the systems. Both kinds of messages had some persuasive effect, but only the moral ones increased people's sense of moral conviction. It also made them less willing to compromise. Reading arguments that revolved around financial concerns actually reduced people's moral conviction. Other work has shown that people with stronger moral convictions use more aggressive bargaining strategies.[14]

What is driving people's unwillingness to compromise along with their rigidity and absolutism? Is it their moral conviction? What would that actually mean? "Moral conviction" could refer to some emotion related to outrage. But if so, why are people experiencing outrage? What aspect of the situation is triggering them? Without an answer to this question, we will never know why people are outraged or how to predict when they will have moral conviction. What does it mean for an emotion to be "moral"? It could mean that it relates to harm. Your feelings about seeing harm come to a small animal are moral because they are a reaction to the pain and suffering of the animal. Yet "moral" could mean something else; there are other theories of morality. According to moral foundations theory, moral senses are triggered in different people by violations of one of seven moral foundations: care, equality, proportionality, loyalty, authority, liberty, and sanctity.[15] So different theorists map feelings to moral convictions in different ways. This makes the notion of moral conviction messy and hard to understand.

There is another possibility, though. Perhaps morality is not what is actually driving the sense of outrage. Notice that what is being violated in each of these cases is a sacred value. Avoiding harming others, loyalty, and all the others are sacred values (they are about action and are treated as more or less absolute). So a different view is that we experience outrage when our sacred values are violated. Obviously this can be described, most of the time, as a moral violation. We often talk about sacred values in terms of right and wrong. But perhaps the appeal to morality is just vague nonsense, unhelpful in the effort to develop a rigorous theory of outrage. Perhaps the real work of determining outrage is grounded in our sacred values. Don't violate my absolute rules of permissible action or I'll be outraged at you! This seems like a more precise account of outrage because sacred values have a more precise definition. Moral prescriptions may serve as a rough guide for behavior, but it is absolute sacred values about action that really lead us.

The idea that sacred values can differ from moral proscriptions implies that there should be cases of nonmoral violations of sacred values that produce outrage. I asked OpenAI's large language model (LLM) ChatGPT (2023 vintage) for an example of one. In response to the prompt, "Give me an example of a nonmoral outrage," it said, "A non-moral outrage could be people being outraged about the design of a newly released smartphone. Perhaps the phone has a feature that is difficult to use or a design flaw that causes it to break easily, and many people are upset and vocal about it. This is not a moral or ethical issue, but it can still generate strong feelings of frustration and anger among consumers." ChatGPT may be wrong; perhaps outrage at bad design is a moral issue in the sense that manufacturers of smartphones are morally obligated to make sure they work. Yet it does not seem moral in the sense of a direct violation of a fundamental moral prescription.

There are other examples of sacred values that appear to have nothing to do with morality. One of my sacred values is that college students should try hard to learn in a classroom setting (I also believe that college instructors should try hard to teach). I don't think that my demand for student effort is a moral demand; after all, the student themself is the one who would benefit. Yet when I encounter students who are slacking, I feel a kind of outrage; I feel that the student is wasting their own time. Here is another example: Some people engage in religious rituals that they consider sacred but are not obviously moral. Is it a moral violation to say God's name in vain? Perhaps, but the only reason would seem to be the belief that God told you not to do it, so the moral argument rests on a belief that many people see as vacuous, and yet even some of those people will avoid saying God's name in vain. Many of us just grew up being told it was wrong and therefore it makes us uncomfortable to do it, even though we do not believe it causes any harm. It is more like wearing a plaid shirt with a clashing shade of plaid pants. It violates aesthetic rules, not moral ones. So although sacred values and moral convictions are highly correlated, they are not the same. Most sacred values are tied closely to moral values. But not all.

Which is the better theory of what determines human intransigence, moral conviction or sacred values? Well, at least the appeal to sacred values offers a definite proposal—that we will be unwilling to compromise if and only if doing so would compromise a rule of action that we claim to be unwilling to trade off. It is not exactly clear what the appeal to moral conviction claims other than a correlation between moral conviction and intransigence—a little bit circular. The important issue is an empirical one: Is there a tendency for people to be unwilling to compromise when faced with a violation of a nonmoral sacred value? The challenge in providing an answer to this question is determining that the sacred value is not moral. You cannot simply ask people because of the ambiguity of the word "moral."

Sacred Repercussions

Marietta presented a couple hundred undergraduates either sacred or nonsacred rhetoric arguing for a specific political position concerning gay marriage, the death penalty, the environment, or gun ownership.[16] The sacred rhetoric was absolutist and did not mention consequences; the nonsacred rhetoric was relativist and all about consequences. Students were then asked to describe the issue, and whether it is the same as most political ones in the sense that it should be decided through the normal democratic politics of discussion and negotiation or, rather, is too important or sacred to be decided by normal democratic politics. For three out of four of the issues, participants in the sacred rhetoric condition were significantly more likely than those in the nonsacred condition to say that it was too important for normal politics. Sacred rhetoric was no more effective than nonsacred rhetoric at changing opinion. But it did change the way students thought about the issue and made them less open to deciding it democratically. This finding, along with the demonstrations I mentioned earlier showing that people with moral convictions on an issue are less willing to compromise, suggests that people hold their own sacred values in such esteem, they favor imposing them on others. When sacred values come into play, it's my way or the highway. Those sacred values that have achieved a broad consensus (thou shalt not murder or thou shalt not steal) can be imposed on others without too much pushback. Imposing sacred values that elicit disagreement, however, requires being more assertive or even aggressive.

Does any of this apply to you or only to others who are less sophisticated? You might want to engage in a little thought experiment at this point. Consider your own view on the current culture wars (wars that might look very different as you read this than they did when I wrote it).

Presumably you hold a position that some reasonable percentage of the population disagrees with deeply, if not violently (on abortion perhaps, or on taxes, war, whether to eat meat, or any other controversial issue). Are you open to having the issue decided democratically? Should we just vote and let the cards fall where they may? Or could that lead to an unacceptable result? Do you believe strongly enough in your position that you think it would be appropriate to force it on society if you could?

I hold views that I would not like to be tested democratically. For instance, I believe that at some point in the near future, only electric cars should be manufactured. I support government mandates to make this happen and do not believe the citizenry should interfere. They just do not know enough about the issue to come to an informed conclusion. My position derives from the belief that humans are responsible for limiting the extent of human-caused climate change. For me, it has become a sacred value. I now value the action of reducing the consumption of fossil fuels so much that I no longer believe in considering other material consequences. My position hardly counts as radical (even if you disagree with it), but even a sacred value reflecting a moderate position can elicit the conviction required to be willing to impose it on others.

So I am someone who does not believe that all policies should be democratically decided and, in the interest of not being a hypocrite, I therefore do not condemn those with intransigent sacred values. I actually admire them. But I am also aware that there are those who hold the opposite opinion to mine with just as much fervor as I have. They believe that government mandates that tell manufacturers what kind of cars they must build violate basic liberties. Manufacturers in a free society have the right to create and sell any product they want that is not designed to kill, maim, or corrupt others. So I admire myself for holding a position that is in direct conflict with that of many others. In that sense, I feed the culture wars.

What about actual war? Do I feel so strongly about my position that I would engage in a physical fight to defend it? Would you engage in a physical fight to defend your position? Or if you don't feel up to fighting yourself, would you support others fighting on your behalf and taking on the good fight? Intransigent sacred values do not cause wars on their own. Thucydides may have been correct when he said that states are driven by three motives: fear, honor, and interest. Only one of these (honor) is a sacred value. Nevertheless, sacred values are inevitably rallied in service of war. The Crusades consisted of eight expeditions over a period of two hundred years in the twelfth and thirteenth centuries. They were essentially attempts by Europeans to grab land in the Middle East. But the reason the expeditions were joined by so many Christians, and the inspiration for their horrific violence and costliness, was their stated objective—to spread Christian religious values and rid the world of infidels. The appeal to sacred values provided a motivation that everyone could understand. It was simple, bare, and absolute. And it appealed to what people cared about the most, the attitudes that gave their communities an identity.

Whether or not wars are the result of conflicting sacred values, a conflict of sacred values results from war. The root cause of war may be a struggle for power, territory, or some other resource, but its instrument is sacred values. Proponents of war inevitably rile up their populations by a call to defend some sacred value or other. The problem is that forcing your view on others leads to a cycle of increasing polarization. French existentialist writer Jean-Paul Sartre expressed the effects of this cycle in a piece in the *Atlantic Monthly* in 1944 that described how the French Resistance reacted to Nazi propaganda in World War II. Here's how the essay begins:

> Never were we freer than under the German occupation. We had lost all our rights, and first of all our right to speak. They insulted us to our faces every day—and we had to hold our tongues. They deported us *en masse*—as workers, as Jews, as political prisoners. Everywhere,—upon the walls, in the press, on the screen,—we found that filthy and insipid image of ourselves which the oppressor wished to present to us. And because of all this, we were free.[17]

The oppressor (German Nazis) painted the French in dark, stark terms, but this just solidified their resistance, increasing their confidence in their own cause. That is how polarization works. One side's vehemence and disdain of the opposition increases the vehemence and disdain on the other side—a positive feedback cycle. Both sides become increasingly certain of their own sacred values and increasingly contemptuous of the enemy's until there are no messy gray areas. There is only the simple black-and-white truth that the enemy is evil and one's own identity is pure—in this case, emancipating the Resistance by reinforcing its sense of free will and strengthening the group's solidarity. This fed many French people's desire and gave them energy to fight.

Many French citizens made an intentional choice to join the Resistance during World War II. Sartre clearly intends to speak for that group. Does the cycle of polarization depend on this kind of intentional choice? Does the dynamic occur when people do not have the explicit goal of supporting one side and opposing the other? It is easier to think about this from the perspective of the French who lived in the part of France that was collaborating with the Nazis (Vichy France). Did they come to align themselves with Germany and share Nazi values only because they were forced to live with them? Did the values of those whom the formerly free French were surrounded by rub off without any real awareness that it was happening? Did they begin to see those who sided with the Resistance as the enemy? Sartre does not offer a direct answer. But it is obvious that people throughout history tend to naturally adopt the lifestyle and values of those around them. So the cycle of polarization does not seem to require people to intentionally choose to adopt a perspective. Perspectives are adopted for us, and cycles of polarization take off when the sacred values held by different sides are in conflict intentionally or unintentionally.

The Ultimate Sacrifice

It does not seem controversial to claim that people, like other animals, benefit from a survival instinct. If there is anything that drives animal behavior, it is the desire to continue living. The instinct to survive explains the extreme lengths people go to when their lives are in peril, like the cannibalism of a fourteen-year-old girl that took place in colonial Jamestown in the desperate winter of 1609.[18] There are evolutionary theories that explain why animals will give up their lives for the sake of their kin, such as to ensure the survival of their genes because genes are shared with kin. Yet evolutionary theory has a much harder time explaining the choice to die under other circumstances, like why people lay down their lives for strangers, or why some people are willing to die for an idea.

History is dotted with examples of people paying the ultimate sacrifice to defend their convictions, from the Zealots of Judaea to philosophers like Socrates, who was forced to drink poisonous hemlock because his views collided with those of other ancient Athenians. Thomas More of sixteenth-century England accepted death rather than give up Catholicism. Japanese kamikaze pilots gave up their lives in suicide dives at Allied warships during World War II. People have strapped on bombs and blown themselves up in an attempt to kill as many people as they can. The modern era includes journalists like Russian Anna Politkovskaya who knowingly assumed risk by reporting on the regime of Vladimir Putin and was then assassinated, or those at the Parisian satirical magazine *Charlie Hebdo*, killed by radical Islamists in 2015 for the periodical's irreverent cartoons of Prophet Muḥammad.[19]

How can we understand such willingness to face death given the strength of the survival instinct? To answer this question, French anthropologist and political scientist Scott Atran has studied in detail one group that is willing to die, terrorists.[20] He concludes that terrorists are driven by sacred values.

In chapter 2, I discussed the social nature of sacred values—that they tend to emerge in communities and provide social identity and group solidarity. One of Atran's central observations is that terrorists are not lone actors who expect to ride into town and save it all on their own. Rather, terrorists are made in small groups. They might have come from a gang of boys who played soccer together when they were kids and grew up with a common sense of victimhood and outrage. Atran reports that two of the young men involved in the suicide bomb attack that killed 1 and injured 40 in Dimona, Israel, in 2008 played on the same Hamas neighborhood soccer team, and so did a number of other suicide bombers. In 2004, 10 bombs exploded on four commuter trains in and around Atocha Station in the center of Madrid, killing 191 and injuring more than 1,800. Later, 7 of the terrorists who plotted the bombings blew themselves up in an apartment in a Madrid suburb after being surrounded by police. Five of the terrorists came from the same primary school in Tétouan, Morocco. Atran concludes, "Taking in fellow travelers and creating a parallel universe devoted to dreams of jihad is commonplace on the road to radicalization."[21]

A key test of whether people are driven by sacred values is whether they are willing to let go of their position for material benefit. Sacred values are absolute and do not permit such trade-offs. In violent conflicts, Atran finds that not only are actors unwilling to trade off their goals for money but asking them to do so makes the situation worse. Financial offers to both Palestinians and Jews to encourage collaboration with the other side made both sides even more disgusted and outraged than they already were. It increased their willingness to use violence too, like suicide bombs.

In stark contrast, a different kind of offer led to more willingness to consider the other side's demands—namely, a symbolic gesture or concession that recognized the other's sacred values. Such a symbolic gesture would include an apology for past behavior. Atran reports examples of such remediating gestures in Indonesia in the conflict between India and Pakistan over Kashmir, and in the negotiations over Iranian nuclear weapons. Iranian leadership is known to have launched a campaign to convince the Iranian population that having nuclear weapons should be a sacred value that stems directly from the country's right to autonomy.[22]

Reframing

Whether or not society can lower the temperature when issues become imbued with sacred values depends on how moored people are in their perspective. Are sacred values set in stone, do they define attitudes toward an issue, or are they a frame that can be substituted with a different frame? Sometimes you look up and see a cloud that looks like a whale. You can see it as a cloud or a whale. You can frame it in both ways. Are sacred values like that? Are they more like a perspective taken in the heat of battle, but one that can be replaced by a less intransigent perspective through some kind of intervention? Mere mortals can only assume a single frame at a time, so it matters a lot. The American Western is a film genre that can be framed as a celebration of rugged individualism and machismo à la Clint Eastwood or as a business model that exploits and reinforces inaccurate stereotypes of Native Americans. Each frame assumes different values and leads to clashing attitudes and beliefs about the genre. Similarly, perhaps political and social issues can be framed according to one or another sacred value, or in a way that motivates open-mindedness and compromise.

Atran offers a number of ideas about how to avoid sacred values frames, thus helping actors to appreciate the complexity of the issues and how limited the actors' understanding is. First, he suggests exploiting ambiguity. Consider a sacred value like "everyone should have equal opportunity." Does that mean that schools and employers shouldn't discriminate against everyone but instead only admit people based on merit? But doesn't "equal opportunity" imply that everyone should get the same education so that they have the same subsequent opportunities? Or perhaps everyone should have the same opportunities from birth regardless of their parents' assets? Perhaps everyone should get the same dietary choices, same soccer clinics, or same bedtime reading? On this interpretation, who your parents are shouldn't matter at all. The point is that the same sacred value has a range of interpretations, and those interpretations vary widely. This is true of many—and perhaps most—sacred values. What exactly does it mean to "keep the Sabbath holy" or "live an honorable life"? These phrases allow a lot of freedom of choice. Some people observe the Sabbath with a short prayer; others devote the entire day to religious services. And this freedom offers a chance to frame an issue in a constructive way. If I encourage someone to live their sacred value of being honorable by being willing to compromise, then I am framing their sacred values in a way that benefits all.

A second idea that Atran offers takes advantage of the fact that people generally have multiple sacred values that prescribe action at different timescales. I might value peace and tranquility today and also value justice in the long term. Different frames will make each value more or less relevant. If I focus on the short term, then my frame will favor promoting peace. If I focus on the long term, in contrast, then my frame will favor promoting justice. If the two sacred values are at odds, as they are whenever injustice prevails, then my framing can make a difference. More generally, different frames can temporarily prioritize different values. Atran observes that fulfilling one value may require delaying others. For instance, prior to the American Civil War of the mid-nineteenth century, President Abraham Lincoln was willing to postpone the emancipation of slaves to save the Union. That particular choice didn't work out so well.

Third, sacred value frames can be deployed to understand an opponent's state of mind and manipulate their behavior. President Richard Nixon's "ping-pong diplomacy" initiative in 1971 had US players go to China to play exhibition matches against the Chinese, the best ping-pong players in the world. The United States was soundly beaten. But losing at ping-pong did not cause riots in the street. The US sense of identity does not depend on its ping-pong performance. It cares more about baseball and bowling. Ping-pong prowess is much more important to China. By allowing China to succeed where it cared, it opened its mind, allowing diplomatic relations with China to open up soon thereafter.

Finally, one of the great diplomatic skills is the art of apology, expressing regret for a harm or violation you take responsibility for (whether or not you actually consider yourself guilty). An effective apology requires that the recipient feel understood, and that whoever is apologizing acknowledges and appreciates the harm or injustice that the victim is experiencing.

It is not good enough to say, "I'm sorry for whatever you think I did." A bankable apology spells out in some detail why the victim feels they were wronged and how they feel about it, such as, "I'm sorry that my action violated this particular sacred value that you hold so deeply, and that you were left with a sense of outrage and despair." You don't need to admit guilt as long as you recognize and acknowledge the other's values.

Sacred Values: The Perfect Rhetorical Weapon

In his book *Moral Tribes*, Harvard psychologist, neuroscientist, and philosopher Joshua Greene identifies what makes sacred values so useful in debate. They give us the power of "heads I win, tails you lose." Generally, it is nice to have evidence on our side, and if such evidence is available, then we will use it. But if I am arguing for gun control and the data show that US states with gun control laws have higher murder rates than states without gun control laws, I can dismiss the data and still argue from my sacred value: People don't have the right to shoot others. And if I am arguing for the freedom to carry weapons and someone points out that the United States has many more mass murders than any other democratic country not at war, I can retreat to my faith in the Second Amendment. Once I have a sacred value in hand, if the evidence supports me, I use it. And if the evidence doesn't support me, I can dismiss it as irrelevant. After all, sacred values require no evidence. Thus sacred values provide what Greene calls "an intellectual free pass."[23]

And because there is always a value—a right or duty—that corresponds to whatever we feel, we always have access to a free pass. If I respond warmly to events at the gay pride parade, I can bathe in the warm glow of people's freedom to express themselves. If I respond with disgust, I can dredge up a biblical passage calling homosexuality an abomination. If I like the government's plan to improve infrastructure, I can think about the great benefits of shared governance. If I hate it, though, I can ruminate about my right to keep the money that I work so hard for. I can almost always find a fundamental right or duty that justifies my reaction, whatever it is. And once I have such a fundamental right, I can pick and choose the data because evidence that does not fit with that right doesn't stand a chance. As Greene says, "Rights are nothing short of brilliant. They allow us to rationalize our gut feelings without doing any additional work."[24]

Everybody wants to feel that their actions are justified. Justification can come from various places. Sometimes we do things just because everybody is doing them (everyone is buying electric cars these days so I think I'll buy one too). Sometimes justification comes from an analysis of what the action will deliver (I think I'll put on my new down jacket today to keep me warm). Other times, justification comes from our sacred values (I'll ride my bike today because I value actions that don't hurt the environment). Appealing to sacred values is more likely when actions are frequent so that we don't have to think through the consequences each time. Appealing to sacred values is also more likely when actions have complicated repercussions. Climate change is incredibly complicated and figuring out the consequences of riding my bike versus, say, taking the bus is above my pay grade. What I can do easily is place a high value on riding my bike to avoid those difficult calculations. Relying on sacred values makes life simpler, especially when those values align with the values of the people around me. The problem is that when they are also at odds with the values of others—who are likely not around me—the clash between my values and yours can spiral out of control. And their apparent simplicity just feeds the fire.

4 Making Decisions by Consequences

I generally suffer from guilt. It is not because of my religious upbringing; my upbringing wasn't so religious. But it does stem from religion. It is because of the tithe. Tithing is the ancient religious custom of giving 10 percent of your income to the church or other religious institutions. I think giving a tenth of your income is a pretty good idea, not necessarily to a religious institution, but to those who need it. One could of course give more. My problem is that I tend to give less. I just do not get around to giving what I think I should and could. One reason for my laziness is that giving is harder than it sounds. In the absence of a religious institution that I want to support, it is not so easy to decide what to give it to. There are so many needy and deserving causes out there.

To solve this problem, a movement emerged during the 2000s that came to be called effective altruism. The movement's priorities include improving global health and development, reducing social inequality and long-term risks to humanity, and increasing animal welfare. The guiding idea is to give to organizations that make the best use of the money.[1] To choose who to give to, effective altruists aim to maximize to help as many people as possible. They ask the question, For every dollar I give, how many people will I help? Answering this question is not trivial. It requires, first, deciding which problems are most pressing. Is it pandemic preparedness, stopping hunger or disease, mitigating climate change or homelessness, or something else? Solving which problems would have the biggest impact? Second, which problems are most tractable? If society will be more successful at preventing climate change than curing cancer, that's a reason to put our resources into climate change. Finally, which problems are neglected? We will make a bigger difference to problems that no one else is attending to than those that others are already working on. The Centre for Effective Altruism points out that Western societies put far more money into counterterrorism than pandemic prevention, yet the coronavirus pandemic killed far more people in a short time than terrorists ever have on US soil. They argue that pandemic prevention deserves a lot more support than it gets because doing so will have a big impact.

Effective altruism was inspired by philosopher Peter Singer, one of the leading proponents of consequentialism. Singer believes that we should make all of our decisions by considering their consequences and choosing options that maximize benefits and minimize costs. He considers questions like the following:

1. "You receive a letter from a reputable international aid organization such as Oxfam. They would like you to donate $500, which they will use to save the lives of several poor African children in desperate need of food and medicine. You feel sorry for these children, but you've had your eye on a stylish new Italian suit, and you'd prefer to save your money for that. You toss the letter in the trash. Are you a horrible person?"
2. "You're walking by a pond one day, when you spot a small child drowning in the water. You could easily wade in and save her, but this would ruin your stylish new Italian suit. So you walk on by. Are you a terrible person?"[2]

If you're like most people (including me), you answered "no" to the first question and "yes" to the second. After all, who doesn't receive letters from charities like Oxfam requesting money on a regular basis and throw them in the trash (well, I recycle them)? And who would not feel responsible for a young child drowning if you were in a position to save them? Singer's point is that the two scenarios are more similar than they seem. In both cases, your action would save the lives of desperate children for the same cost. In fact, you would be saving more lives in the first case so there is even more reason to act.

Singer takes an extreme consequentialist position. He contends that the nature of the action you take does not matter at all. It is the result that matters. For many decisions, it is hard to argue against consequentialism. If you are making dinner, you should choose the recipe that leads to the tastiest and healthiest dish with the least cost and effort. If you are choosing a medical treatment, you should choose the most effective one with the fewest side effects that you can reasonably afford. The list of decisions that invite consequentialist solutions is long.

Don't be fooled into thinking that optimizing consequences is easy. Optimization requires its own kind of values—not sacred values but values on outcomes. Which outcomes should we prefer? The effective altruists believe that charity does not begin at home but instead where we can help the most. By donating $100 to a charity in the United States or some other rich Western country, you might feed a small family for a week if you're lucky. For the same amount, you can feed at least three such families in the most poverty-stricken parts of the world. So if you are trying to save the most people, more charity should go to poor countries where hundreds of millions of people live at subsistence levels of income rather than to rich countries even in the likely event that you live in one. Effective altruists believe that transferring resources to poor countries can do more good than keeping it all at home. In fact, they believe in extending the "circle of concern" to nonhumans. They support not only charities that provide for people but those that improve the quality of life of farm animals too. This desire to spend resources to prevent the suffering of farm animals is of course not universal. Consequentialism involves a lot of choices.

Consequentialism is hard in another way as well. Decisions never guarantee an outcome. The best you can do is make an outcome more probable. If you choose to spend your spare income by sending it to a charity feeding those suffering from drought and war in Sudan, you are increasing the probability of saving lives in Africa. But you cannot know for sure that the money will get there, or that it won't be stolen by corrupt mercenaries or be spent bribing government officials. Your charitable donation has a range of possible consequences and you cannot be certain which will occur. Even simple, everyday decisions have uncertainty. If you choose Coke over Pepsi, there is some chance that what you get is too warm or has gone flat. There are few guarantees in life (except death and taxes). When evaluating charities, effective altruists talk not about saving lives but instead about the expected number of lives saved. Figuring out consequences is hard.

The Two Pillars of Consequentialism

Decision making involves selecting an option from a set of possibilities. To make a decision based on outcomes, you need to solve two problems.

The first problem concerns *belief*. What do you believe the effects will be of the choice you make? Answering this question requires knowing a little about how the world works, how the world will turn your action of choosing into a set of consequences. Your choice is more likely to cause some outcomes than others, and it is your beliefs about how the world works that will tell you which outcomes are more or less probable. Beliefs are not entirely about you; they are about facts and what leads to change in the world. If I believe that turning a screw clockwise will tighten it (righty-tighty and lefty-loosey), then I will choose to tighten a screw that way. But if you believe righty-tighty means turning counterclockwise, then you will tighten in the opposite direction. Your beliefs about how the world works determine the option you choose.

The second requirement for consequentialist decision making is that you know how much you *value* the possible outcomes. These are not sacred values; they are values about outcomes, not actions. You might prefer that a screw is tight rather than loose, that the sun is shining rather than that the sky is cloudy, or that your baby is sleeping rather than crying. In each case, you place more value on the first outcome than the second.

In order to make a decision, you have to know what you want—which outcomes you prefer over others. Otherwise your decision has no guide. Outcome values reflect your own individual preferences. In that sense, values are about you. They concern what you want or what you like (as we will see, these are not the same).

Both beliefs and values contribute to the difficulty of consequentialist decision making. We use beliefs to calculate the likely consequences of decisions. This can be hard because the necessary calculations can be complex. They might require difficult causal reasoning (e.g., What would be the consequences of the United States invading China? What would be the outcomes of giving all of your money to a worthwhile cause? What would be the consequence of adding a little pineapple juice to your soufflé?). The complex calculations might be mathematical (e.g., Will it take more energy to get to my destination if I travel fast on the highway or meander along on backroads?). Most often, what makes it hard to calculate outcomes is uncertainty. The result of a decision almost always depends on events outside our control, so the best we can do is guess the probability of an outcome.

Judging probability is a challenge because we are rarely in a position to compute it. If I choose to move to Colorado rather than Wisconsin because I like the mountains and climate of Colorado, the outcome depends on events whose probability I cannot know. Will I get along with my neighbors? Will crime or traffic where I plan to live become unmanageable over the years? Uncertainty is often impossible to calculate even for simple decisions. I might choose to take a sip of my beer, but what if an earthquake occurs at that moment and spills it all over me? What is the probability of an earthquake? Fortunately, this is an uncertainty you can get away with neglecting. Still, determining consequences and their probabilities poses a number of challenges.

Beliefs are not the only source of difficulty in decision making. Values may be even more of a problem for everyday decisions. We frequently do not know what we want. Do I feel like eating ravioli or summer sausage? Am I in the mood to ride my bike or walk? Do I really like sharing my life with this other individual? Our desires are just not always clear to us and that can produce conflict. Sometimes soul-searching helps. We might be reminded how much we regretted ordering ravioli last time. But sometimes deep reflection just reveals more costs and benefits of each option and increases our conflict. Much of the psychology of decision making shows we are more malleable than we appear. Whether we want to have a beer or not turns out to depend on how the options are described to us, how we are asked to choose, what our goal is at the moment, and who is around us when we are deciding. If everybody else is having a beer (or smoking or eating too much or laughing), then we are more likely to do so too.

So consequentialist decision making entails difficult external assessments based on beliefs about how the world works as well as difficult internal assessments of what outcomes we prefer. How can we make decisions in a way that respects both of these considerations? The great French thinker (and former child prodigy) Blaise Pascal provided an early hint at an answer in his *Pensées*, published after his death in 1662. As a seventeenth-century European, Pascal was faced with the burning question of whether or not God exists. He concluded that the question was unanswerable on the basis of reason alone, but that an even more important question could be answered: whether or not we should act *as if* God exists. Consider the following table, with blank spots for describing the consequences of acting as if God exists under two states of the world—that God does exist and that God does not exist.

	States	
Action	God exists	God doesn't exist
As if God exists		
As if God doesn't exist		

How would you fill in this table? What are the relevant consequences? If you lived in France in the seventeenth century, you might fill it in something like this:

	States	
Action	God exists	God doesn't exist
As if God exists	Minor frustration and rule following on earth followed by eternal bliss in heaven	Minor frustration and rule following on earth
As if God doesn't exist	Minor pleasure and freedom on earth followed by eternal hell and damnation	Minor pleasure and freedom on earth

With this little table, Pascal laid the groundwork for consequentialist decision theory. The table encodes both beliefs (which states of the world are relevant to the decision) and values (the costs and benefits that apply to the consequences of taking each action depending on the true state of the world).

The table leaves out the probability of each state—that is, the probability that God exists. Determining that probability is tough and exactly what Pascal wanted to avoid because he thought it could not be reasoned about. Also missing is an explicit statement of values. How much do we care about minor frustrations versus eternal bliss? In seventeenth-century France, the answer was obvious, and it is for many people today. Eternal bliss trumps minor frustrations by an infinite amount, and hell and damnation are infinitely bad. In other words, the consequences are so extreme if God exists and so piddling if God does not exist that we should act as if God exists.

The pleasure obtained if God does exist is so positive and the potential suffering if we are damned so terrible that the likelihood that God does in fact exist does not even matter. The probability is trumped by the consequences. Based on this, Pascal made a bet that God exists by acting as if God exists and recommended that everyone else do so as well. That is why this analysis is called Pascal's wager.

Getting the Most Out of Your Decisions

Implicit in Pascal's wager is the idea that we want to get the best consequences when we make a decision. We want to maximize; we want to get the highest value. But the highest value of what? If we are investing for retirement, presumably we want the most money. And if we are making decisions to complete a crossword puzzle, then presumably we want the most answers correct. So the quantity we are maximizing depends on what we are doing. Pascal was focused on eternity—heaven and hell—where it seems unlikely that either money or the number of words correct matter. What did Pascal want us to maximize?

An answer was given by another former child prodigy, nineteenth-century English philosopher John Stuart Mill. Mill proposed that when we are making a decision, we should try to maximize happiness. He called this the greatest happiness principle. "Actions are right in proportion as they tend to promote happiness, wrong in proportion as they tend to produce the reverse of happiness." But what is happiness? According to Mill, "By 'happiness' is intended pleasure, and the absence of pain."[3] Every decision we make, every action we take, is for the purpose of obtaining pleasure or avoiding pain. This squares nicely with Pascal. Heaven does seem to be a good place to go for infinitely enduring pleasure, and hell is the right choice to suffer eternal pain.

The kind of happiness that utilitarians like Mill and Singer subscribe to is not a selfish personal happiness but rather the common good, everybody's happiness. They are actually quite extreme about it. The great philosopher Jeremy Bentham's rule was, "Everybody to count for one, nobody for more than one."[4] According to this dictum, even the decision maker does not count more than anyone else. This is sometimes called the principle of equal consideration. All decisions should be aimed at maximizing everyone's happiness, with the decision maker being only one of those people.

This has some striking implications. For instance, it becomes incumbent on us to severely harm or even kill innocent people (including ourselves) if doing so will improve the common good.

Oxford philosopher Guy Kahane and his colleagues point out that this principle makes both a positive and negative demand.[5] They call the positive dimension "impartial beneficence." It states that we should be willing to sacrifice ourselves for the sake of others. This is behind the effective altruists' call to take positive action and be magnanimous and charitable, especially to the most impoverished. They call the negative dimension "instrumental harm." It requires that we are willing to sacrifice the well-being of others for the greater good. To illustrate, Singer has gone so far as to argue—to the dismay of many—that the infanticide of severely disabled babies who lack rationality, autonomy, and self-consciousness can be justified, as in cases of spina bifida or anencephaly. Singer believes that such cases involve intrinsic suffering and it is appropriate to reduce suffering. On Kahane's analysis, Singer is calling for an action that can only be characterized as harmful or negative with a consequentialist justification.[6]

Kahane and his colleagues have shown that impartial beneficence and instrumental harm provide mostly independent descriptions of laypeople's intuitions about right and wrong. Some people subscribe to impartial beneficence, others to instrumental harm, some to both, and some to neither. Philosophers, however, differ from laypeople in that they tend to align on the positive and negative dimensions, either accepting or rejecting both demands. This is likely because philosophers have thought about utilitarianism and made a conscious decision whether they are for or against it. Laypeople, in contrast, just have their intuitions, and intuitions do not necessarily line up with any philosophical theory.

The greatest happiness principle has not stood the test of time as a description of how people respond to consequences. Too many decisions seem to be about something other than pleasure or pain. If we always buy the cheapest brand of dishwashing liquid when we go to the supermarket, does that increase pleasure or decrease pain? It may just maximize the amount of money in our bank account—a quantity that for many has emotional valence only when it is very low or very high. What about decisions that decrease pleasure and increase pain, like those made by soldiers who join an attack where they have a good chance of being maimed or killed, or the choice of a boxer to enter the ring to likely get pummeled by an opponent?

One could argue that each of these actions leads to the pleasure of battle, happiness of others, or potential rewards that come with success, but this would be to decide what constitutes pleasure after we already know the decision, so the notion of pleasure isn't helping understand why the person did what they did. It is merely making us struggle to understand what pleasure is.

A deeper issue with the greatest happiness principle is that happiness turns out to be multiply ambiguous. First, what is it that makes us happy? Is it getting what we want—satisfying our desires—or what we like? It turns out that what we want is quite different from what we like. This is the conclusion of research by Kent Berridge of the University of Michigan. Wanting refers to desire. It is measured by the tendency to approach whatever it is we want or to choose it. Liking, in contrast, is a positive response once we have the object in hand. For instance, infant humans, chimps, and rats all protrude their tongues when they like something that is in their mouths. So tongue protrusion is one measure of liking. Berridge has shown that wanting and liking are mediated by separate brain systems in mice and rats, and the same disassociation likely holds in humans.[7]

One study showed people sequences of TV advertisements of over a thousand objects and measured how much participants liked each ad as well as how much they wanted the advertised object.[8] Although people liked the ads the first time they saw them, their liking rapidly declined so that they liked them much less the second through sixth time that they saw them. Who wants to see the same ad again and again? But the pattern for wanting was completely different. People wanted the object more and more the more times they saw the ad. Mere exposure to the object seemed to increase desire for it.[9]

The fact that wanting and liking are so different has critical implications for how we make decisions. For one, it means that there should be cases of liking but not wanting something. Most of us have had this experience in, say, art museums where we might like a piece of art without wanting it to take over our living rooms (I feel much the same way about Notre-Dame Cathedral). It also means that there should be cases of wanting without liking. Sadly, this is common in cases of addiction. Those suffering from substance abuse, gambling addiction, and other kinds of addiction frequently report that even though they strongly crave doing the addicted activity, they do not enjoy it once they are doing it. This is one of the sad facts of life. The things we want most tend to disappoint.

Happiness is ambiguous in a second sense as well, best described by Nobel Prize–winner and psychologist Danny Kahneman. He points out that decisions are made with the desire to get the optimal experience, or what he calls "experienced happiness." But what actually drives decisions is our memory for what we experienced in the past (called remembered happiness) or what we expect will provide the optimal experience in the future (called prospective happiness).[10] If we're choosing between two dishes at a restaurant, then our choice is likely to be based on our memory for similar dishes that we have had previously. After all, there is little other basis to judge. And if choosing between two possible vacation locations that we have never been to, then our choice is likely to be based on predictions we make about what our experience will be like in each location—again a reasonable strategy. Neither strategy, though, targets exactly what we really care about and what the actual experience will be in each case.

The fact that the measures of happiness that we use to make decisions, remembered and prospective, differ from the measure of happiness that we want to maximize when we make a decision, experienced happiness, is a problem. It introduces a systematic bias into our decision making because our memories and predictions differ systematically from actual experience. Unlike actual experience, memories and predictions depend on the stories we tell ourselves. My memory for a dish at a restaurant will be biased in favor of how I felt when I ate it—how good the company was, and whether or not my ex was sitting at the table beside me. These are factors that are not relevant to whether the dish would be enjoyable at a different time and place. My prediction about a vacation will be biased by stories that I have heard. If my friend tells me that they lost their room key at a resort in South Carolina and ended up having a great time drinking the night away in song with some Australians with a ukulele, I will have a good feeling about that resort even though one could meet singing Australians anywhere (but only if you are lucky). I should be choosing based on the quality of the distinctive aspects of the location, but instead I imagine a scenario that conforms to the stories I tell myself. In general, the considerations that guide my decision making are not perfectly aligned with the experience that actually follows from my decision.

This distinction hurts the most when we are making decisions about social policy in domains rich with stories. Debates about whether or not to support a welfare bill tend to be governed by anecdotes relating cases of welfare getting people out of poverty or, alternatively, causing someone to become dependent on government handouts. The effects of welfare are actually exceedingly complicated, and the consequences that arise cannot be reduced to simple stories.

Despite Mill's bona fides, happiness is thus a problematic guide to decision making because the kind of happiness that we want to maximize when making a decision, experienced happiness, is a different animal than the kind of happiness we have access to when deciding, remembered or prospective happiness.

Given that happiness does not serve as a perfect currency for decision making, is there an alternative? What else might we maximize to make decisions? One proposal is that rather than striving for happiness, we should strive for meaning. This is the answer that Viktor Frankl came to after three years in a Nazi concentration camp and losing almost his entire family to genocide. He concluded that what drives people to go on living is the search for meaning, to feel that one's life has a purpose.[11] This quest for significance applies to decisions small and large. A decision to applaud a performance when one is a member of a large audience has negligible consequences as one person's applause is overwhelmed by the crowd. Nevertheless, it signals to the person applauding their appreciation for the performance. Other decisions involve massive consequences and appear to be driven by a desire for one's life to have meaning too. Psychologist Arie Kruglanski has studied terrorists and argues that what drives violent extremism is a quest for personal significance.[12] Suicide bombers' actions are the only means at their disposal to give meaning to their lives—a sense of significance that arises from the feeling that they have measured up to the values they share with those they are closest to.[13]

The Arithmetic of Utilitarian Decision Making

Utilitarianism came into its own as a rigorous theory of decision making when mathematician John von Neumann joined forces with economist Oskar Morgenstern. They set out a series of basic assumptions called axioms that a rational decision maker would have to follow and proved that following those axioms entails always choosing by utility (maximizing expected utility to be precise). I will not go through their proof, but I will give you an example of a rational axiom so you can get an idea of what I'm talking about.

One axiom is called transitivity. It states that if you prefer option X to option Y, and you also prefer option Y to option Z, then you should prefer option X to option Z. Mostly, this is sensible. If I choose vanilla ice cream over chocolate and chocolate over strawberry, then it would be weird if I chose strawberry over vanilla. There are rare circumstances where transitivity is violated. In sports, say, there are cases where a team regularly beats a second team and the second team beats a third team. Yet the third team has some surprising advantage over the first team and regularly beats it. If you see such a rare circumstance, you should bet on teams in a way that violates transitivity and you may well make a lot of money. That is because the situation is so rare.

Most of von Neumann and Morgenstern's other axioms also concerned preferences because preferences dictate choice: We choose the option that we most prefer. Their key idea was to associate options with a number that they called utility (U), so their theory is a type of utility theory. If I prefer X to Y, then the utility of X to me is greater than the utility of Y, $U(X) > U(Y)$. Normally, we assume that having more money is better than having less. This implies that the utility of a larger amount of money is greater than the utility of a smaller amount. But not all differences are necessarily equal. Obtaining \$100 if you have nothing means a lot more to you than obtaining \$100 if you already have a \$1,000. If you don't have anything, then \$100 means you can buy lunch. If you already have \$1,000, then you can buy lunch anyway. We can describe this pattern of utilities in the following way using what is called a utility function:

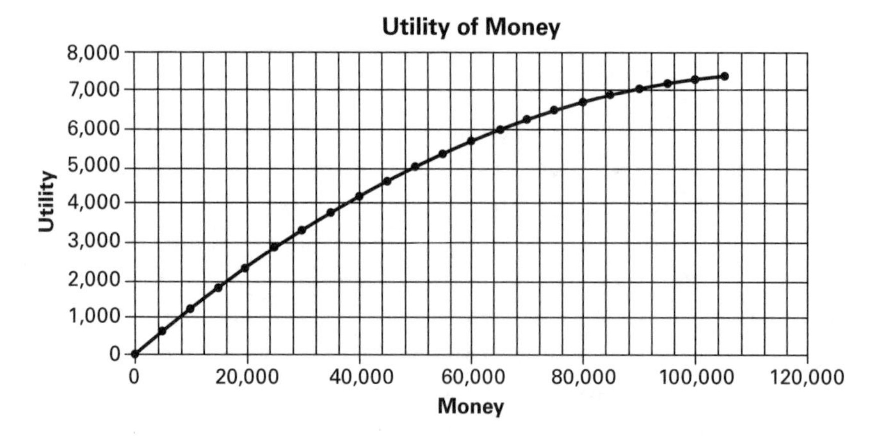

Figure 4.1

You can see in this curve that the extra utility you gain (on the y-axis) from an additional $100 depends where you are on the x-axis. The further you are to the right, the less of a utility bump you get.

Not all goods behave this way. The beauty of the concept of utility is that I can talk about the utility of anything—the utility of going dancing, going to church, or having a good cry. Utilities are helpful because they allow us to talk about any decision in quantitative terms. You can even talk about the utility of goats. You might argue that having a goat is a good thing and has positive utility (although this is much less clear if you live in a city). You might even assert that having two goats is better than having one. But with goats, there is a definite limit on the number you want to have; more is not necessarily better. So the utility function for goats is not constantly increasing but instead starts descending. For city dwellers, it starts descending at the beginning.

The power of von Neumann and Morgenstern's theory is that they provided a way to handle uncertainty. Let's say I do decide to get a goat. At the time of purchase, this could lead to a host of outcomes. I could fall in love with it and live happily ever after. It might eat all of my furniture. Or it might just eat some of my furniture. I really don't know what's going to happen. I am chock-full of uncertainty. Von Neumann and Morgenstern handled the uncertainty in the same way that most people do these days. They used probability. My task as a utility theorist is to assign a probability to each outcome of getting a goat. I put the probability of living happily ever after very low. I'll give it a 0.1. I also don't think it's likely to eat all of my furniture because I can close some doors so it can't get at some of it, so I'll make that low too, again a 0.1. Yet I deem the probability pretty high that it will eat some of my furniture, so a 0.8. In fact, 0.8 is predetermined because the probabilities of all outcomes of a single option have to sum to 1 ($1 - 0.1 - 0.1 = 0.8$).

We can already see that von Neumann and Morgenstern's utility theory represents both components of a decision, belief and value. The beliefs are encoded as probabilities (what I think will be the consequences of each option), and the values are encoded as utilities (a numerical representation of my preferences). The mathematical theory makes use of one more idea, expectation.

The expectation of a random variable is what it sounds like, what we expect the value to be of a variable that takes on different values in a way that we cannot completely predict (whatever determines the variable's value has, from the decision maker's perspective, a random component). The expectation is a kind of average. Sometimes I get 5 emails in a day, sometimes I get 20, and sometimes I get 100. The average is $(5 + 20 + 100)/3 = 41.67$. There are a lot more days, however, when I get 20 emails than days when I get either 5 or 100. I would estimate the probability of getting 20 emails to be 0.7 (on 7 of every 10 days I get 20 emails), the probability of getting 5 to be 0.1, and the probability of getting 100 to be 0.2. The average does not capture the number of emails I expect because most days I get 20. Instead, I calculate the expected value, an average weighted by the probabilities of each outcome, $0.1 \times 5 + 0.7 \times 20 + 0.2 \times 100 = 34.5$. I should expect about 35 emails per day.

Because von Neumann and Morgenstern's theory requires you to calculate the expectation of the utilities you assign to each outcome of each option, it is called expected utility theory (hereafter EU theory). Let's say you're deciding whether or not to buy the following state lottery ticket for $1:

Dewey, Cheatem, and Howe State Lottery

With probability	You win
0.6698999	$0
0.2	$1
0.1	$2
0.02	$5
0.01	$10
0.0001	$100
0.0000001	$1,000,000

What is the expected value of this lottery ticket? It is $0.67 \times \$0 + 0.2 \times \$1 + 0.1 \times \$2 + 0.02 \times \$5 + 0.01 \times \$10 + 0.0001 \times \$100 + 0.0000001 \times \$1{,}000{,}000$ with a final value of 0.71 or 71¢. That is less than $1 so the ticket is not worth $1. But EU theory says that the money values we might win should not determine the value of the ticket to us. What matters is the utilities of those values. In the next table, I have assigned a set of potential utility values that represent the preferences of someone who really likes large amounts of money—the larger the better. For this person, winning the jackpot is huge, worth 5,000,000 in utility, over 5,000,000 times the value of spending $1 (which is worth only 0.8 in utility).

With probability	You win	U($)
0.6698999	$0	0
0.2	$1	0.8
0.1	$2	1.8
0.02	$5	5
0.01	$10	12
0.0001	$100	200
0.0000001	$1,000,000	5,000,000

Now that we have utilities, we can calculate expected utilities ($0.67 \times 0 + 0.2 \times 0.8 + 0.1 \times 1.8 + \ldots + 0.0000001 \times 5{,}000{,}000$). This comes out to 1.08. The expected utility of this lottery ticket is 1.08, whereas the expected utility of $1 is $1 \times 0.8 = 0.8$. The law of expected utility is to choose the alternative with the most expected utility. So to a person with the utility function shown, buying the lottery ticket is well justified.

Risky Business

Imagine I offer the following attractive option to a group of people. I will either give them $100 or they can choose a risky bet. The bet is that we will flip a fair coin and if the coin comes up heads, I give the person $200, but if it comes up tails, they get nothing. There are three kinds of people. One kind of person refuses the bet and takes the sure thing. According to EU theory, such a person has a utility function that gives a higher expected utility to $100 for sure than to even odds of getting $200 or nothing. For this person, $U(\$100) > 0.5 \times U(\$200) + 0.5 \times U(\$0)$. We call such a person risk averse because they would prefer a sure thing to the possibility of a larger amount when the expected values are the same. It turns out that if your utility function has the downward-curving (concave) shape of the utility function shown earlier, that implies that you are risk averse.

A risk-seeking person chooses the bet. According to EU theory, for that person, $U(\$100) < 0.5 \times U(\$0) + 0.5 \times U(\$200)$. If they have an upward-curving utility function (convex), like the lottery ticket buyer described in the previous section, then they will be risk seeking.

Finally, a risk neutral individual is indifferent between the two options. Such a person does not know what to choose. The expected values of the two options are the same in this case ($100). Because a risk neutral person's utility function is a straight line relating money to utility, they are essentially choosing according to expected value. The utility of money is just the value of the money to such a person.

When we do the analysis this way, each person is obeying the law of expected utility. Each of them is choosing the alternative with the most expected utility. They just have different utilities and hence make different choices.

Two-Envelope Problem

For the sake of completeness, I do need to point out that not everything about EU theory is obviously kosher. Problems with the coherence of the theory have been pointed out. I will note one simple problem just to give you a taste of some of the academic discussion of the issue. This one is not just a problem for EU theory but rather for any theory that relies on calculating expectation. Let's say I have two envelopes and tell you the following (true) facts about them:

- they each contain money
- one contains twice as much money as the other
- looking at and feeling them gives you no hint about which envelope has the larger amount
- you get to choose one and keep it

Great. Say you choose the one on the right. What is its expected value? Who knows? It is whatever amount of money is in it. Let's call that amount M. Now here is a simple question. What is the expected value of the other envelope? Well, we know that one envelope has twice the amount of the other, so it either has $0.5M$ or $2M$. What is the probability of each state of affairs? Well, any reason you might have for thinking it is $0.5M$ also applies to thinking it is $2M$. The two possibilities seem utterly symmetrical, so they would seem to have an equal likelihood of 0.5. Thus the expected value of the other envelope is $0.5 \times 0.5M + 0.5 \times 2M = 1.25M$. The expected value of the envelope in your hand is *M*, but the expected value of the other envelope appears to be 1.25*M*! If I offer to trade with you, should you accept? According to EU theory, the answer is yes because you are supposed to choose the option with the higher expected utility, and presumably the envelope with the higher expected value also has a higher expected utility. In sum, you randomly chose an envelope, and regardless of your choice, you should now choose the other envelope. Odd, no? But it is worse than that because now that you have traded, you have the other envelope in your hand. Should you trade back for your initial choice? Well, if you assume that the envelope in your hand has a fixed amount of money and then calculate the expected value of the other envelope, the answer is yes. You should trade back. And you should do so forever.

Clearly, something must be wrong with this line of reasoning. But what? Decision theorists and philosophers have proposed a variety of solutions to this paradox. It is only one simple paradox in the world of decision theory, though. I think it should at least raise questions about whether EU theory is the way we should always be making decisions.

Reasons to Obey the Law

Let's get back to the strengths of EU theory. To repeat, the law of expected utility is to choose the alternative with the most expected utility. Here is a good reason to obey this law. If you do for every decision you make, you will get the most utility possible over the course of your lifetime. This is good to the extent that utility represents how much you really want things. Because then, if you choose by the law of expected utility, you will get the most of what you want. What could be better than that?

There is another reason to choose according to the law of expected utility. Because EU theory is equivalent to the axioms of rational preference, if you violate the theory, then you are violating at least one of the axioms. And if you violate any of the axioms, you can be turned into a money pump; I can extract all of your assets from you. To see how this works, let's consider transitivity again. Say I have three cars, an Audi, Mercedes-Benz, and Chrysler. If you have a transitive preference ordering, then it might look like this:

Transitive: Audi is preferred to Benz, Benz is preferred to Chrysler, and Audi is preferred to Chrysler.

But if your preference ordering is intransitive, it could look like this:

Intransitive: Audi is preferred to Benz, Benz is preferred to Chrysler, but Chrysler is preferred to Audi.

Let's see what I can do to you if you have the intransitive ordering. The first step is generous of me. I give you the Benz. Now I have the Audi and Chrysler, and you have the Benz. But you prefer the Audi to the Benz, so you should be willing to trade for a small fee. Maybe you would be willing to trade for $1,000, but maybe you would play hardball and only give me $1 to make the trade. That's fine. Let's trade. Now you have the Audi, and I have the Benz, Chrysler, and one of your dollars.

But wait. You prefer the Chrysler to the Audi. So let's trade (for a mere $1). Now you have the Chrysler, and I have the Audi and Benz plus two of your dollars.

But wait. You prefer the Benz to the Chrysler. So let's trade for my minimal $1 fee. Now you have the Benz, and I have the Audi and Chrysler plus $3 from you. Haven't we been here before? This is where we started except I am $3 richer and you are $3 poorer. As long as we are making our decisions based only on expected utility, we can keep going. In theory, the trades should never stop until you no longer have any dollars to give me.

This is one sense that being intransitive is irrational. The fact that people would see the trick in real life and not fall for it misses the point. The point is that it is possible to trick you unless your preferences are consistent with the axioms of EU theory. If you are an expected utility maximizer, that is enough to avoid irrational choices.

Depicting Decisions with Trees

When faced with a decision with multiple options and outcomes, if one wants to go the consequentialist route, it can be helpful to draw decisions out with a tree structure that represents the probabilities and utilities of each option. I present a small concrete example here as a practical lesson about how you can deploy EU theory to make actual decisions that you face. The method I am about to describe takes some time and effort, and so is only meant to be applied to difficult decisions that are important enough to warrant the time and effort.

Imagine that your alarm clock rings and wakes you up. You are now faced with a decision. Should you (1) shut it off and get out of bed, (2) shut it off and go back to sleep, or (3) hit snooze and go back to sleep? Let's say that you have a job and it is a workday. So the consequences we will consider that are associated with option 1 are:

- you get dressed, go to work, and have a dreary day
- you get dressed, go to work, and have a fabulous day
- you get dressed, go to work, and find that your boss has decided to fire you

The likely consequences associated with option 2 (going back to sleep) are:

- your boss decides to fire you
- nobody misses you, and you catch up on your sleep

Option 3 (hitting snooze) has the same likely consequences as option 1 except that you will arrive at work late.

We can depict all of this with the following decision tree. The tree starts with a square or decision node on the far left. The lines (edges) coming out of it represent your options. There are three of them (get up, sleep in, or snooze). You have control over which of these edges you follow.

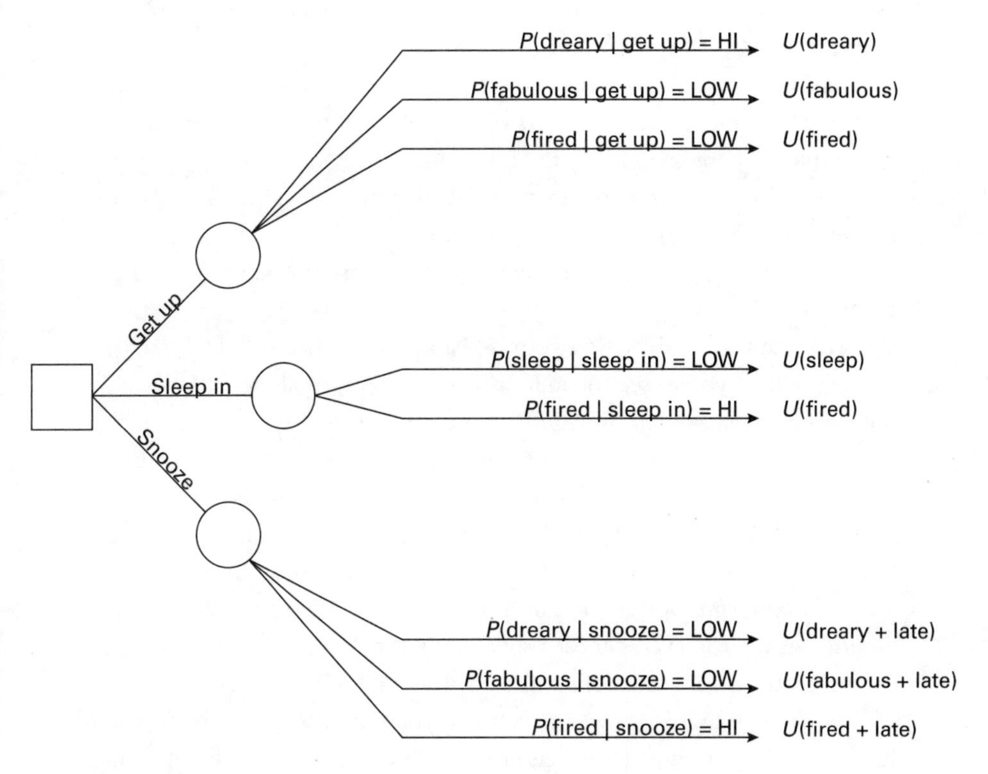

Figure 4.2

Each option is associated with a chance node represented by a circle in the figure. Each edge coming out of a chance node represents a possible outcome. The world external to you determines which of these edges gets followed. Each option leads to different possible outcomes. They are labeled with conditional probabilities. That is, conditional on you choosing one of the options, there is a probability of a particular outcome. For example, given that you get up, the probability of you having a dreary day is written as *P*(dreary | get up). The variable on the right side of the vertical bar represents the conditions that we are assuming to be true or given (that you have gotten up). The variables on the left side of the bar represent the events whose probability we are judging (the probability of having a dreary day assuming that we have gotten up). Given a particular option, the probabilities of all the outcomes have to sum to one because each outcome is mutually exclusive of the others and together they are exhaustive of the possible outcomes. You have to judge these probabilities based on your beliefs about the world. In this case, I imagine that *P*(fired | get up) is lower than *P*(fired | snooze). Your boss is more likely to fire you if you snooze and arrive late than if you arrive at work on time.

Finally, each final outcome node is represented by the utility of that outcome. These *U* values are up to you, but should reflect how much you prefer that outcome relative to the other ones.

Once you have assigned values to all of these probabilities and utilities, you can determine the expected utility of each option in the way described above. Then the final step is to choose the option with the highest expected utility.

The advantage of depicting your decision using a decision tree like this is that it allows you to see the main issues (the options along with possible outcomes as well as their probabilities and utilities) all in one place. Engaging in this depiction exercise can be helpful for decision making because it forces you to make explicit all the considerations related to consequentialist choice and allows you to see the decision as a whole. But it is not a silver bullet because if you already know where you want to end up, you can always choose your options and utilities to give you the conclusion you are aiming for. You can even choose probabilities that are in your favor to some extent, although they are constrained by the way the world is.

My advice, for what it's worth, is that if you have an important decision to make, going through an exercise like this is worthwhile. It allows you to see how robust your decision is from a consequentialist perspective. That is, you can play with the probabilities and utilities, and see if changing them makes a difference. If not, you have probably arrived at the best decision. But once you have done the analysis and thought about all the issues carefully, then you should rely on your intuition to make the final call. As we will see, intuition is able to view situations more holistically—less selectively—than this kind of careful, deliberative analysis.

5 The Art of Belief: How People Think About What Is

In 9 CE, Arminius, chieftain of the Germanic Cherusci tribe and leader of an alliance of Germanic tribes, led a surprise attack on a Roman force led by General Publius Quinctilius Varus called the Battle of the Teutoburg Forest. The Romans comprised three legions along with some auxiliary forces and cavalry, around twenty-five thousand soldiers, or about one-fifth of the entire Roman fighting force. The Germans attacked at a place now called Kalkriese, where the Romans were forced by the landscape (a single road through a forest) to spread thinly into a long column of perhaps twenty kilometers. The initial attack routed the Romans. But there were enough survivors to attempt an escape. The clever Germans dug a trench that hemmed the escaping Romans in a small gap between a hill and great swamp. Arminius's forces then attacked again from behind a wall they had built along the road, slaughtering the Romans. They left fifteen to twenty thousand dead. Historians credit this military rout for Rome's failure to ever conquer German lands beyond the Rhine.

Why did Arminius decide to attack at Kalkriese? It turns out that Arminius had a big advantage over the Romans. He knew Roman strategy and tactics. He had been taken hostage by the Romans at the age of ten and brought up like a Roman aristocrat, serving in the Roman military and assisting Varus himself. Arminius eventually became a commander of the Roman auxiliary units. So he had a deep understanding of the Roman military. He knew that the Romans would be at a disadvantage when stretched thin, attacking from behind a wall would reduce the impact of the Romans' stronger weaponry, and the element of surprise was critical. He was reasoning in true consequentialist fashion, based on beliefs about what his opponent Varus believed. Specifically, he believed that Varus did not expect the Germans to attack at Kalkriese.

All of these beliefs provided the framework that encouraged Arminius to attack when he did, thus determining the course of European history. The Germans certainly took a risk by attacking. They could not be sure of the outcome before it happened. But their beliefs made the probability of success high enough that they took the chance. They succeeded.

The philosophers among you will have noted that I am not distinguishing knowledge from beliefs, a celebrated distinction in philosophy. I take the standard position among cognitive scientists and psychologists that both knowledge and belief refer to mental representations—structures in the mind that contain information about categories and individual objects in the world, including a reference to them. The human mind treats the representations that correspond to knowledge and beliefs identically. Knowledge and belief may differ in their truth value or justification, but I see no reason to think they are processed differently by the mind. It is true that there are certain articles of faith that we refer to as *belief* and not *knowledge* in English (e.g., belief in a deity who controls affairs), yet this does not change how to understand the role of mental representations in decision making.

Mental Representation and Decision Making

At the time of this writing, LLMs like OpenAI's GPT-4 are taking the world by storm. By the time you read this, they will be commonplace. They are huge neural networks—computational devices composed (again at the time of writing) of trillions of connections between incredibly simple processing units, trained on fantastically huge datasets whose size approaches that of the entire internet. Today, their capabilities are mind-blowing. They pass bar exams; they can identify the objects in photographs better than people can; they can compose poetry and write computer programs; and they can correctly answer difficult questions about cause and effect. Progress is so fast that I cannot even imagine what they will be capable of when you are reading this.

What I find most striking about LLMs is how humanlike they are. LLM networks are brain-like in one sense—they involve complicated interactions among billions of relatively simple processing units. They are not like the brain in other ways. But they are like the mind. That is, they operate according to the same abstract principles as human memory.

Cognitive psychologists think about memory as comprising three stages: encoding, storage, and retrieval. At each stage, LLMs are like humans. Information is encoded according to how surprising it is and represented in a way that conforms to what the system already knows. Information is stored in a way that is distributed across trillions of connections. Each connection participates in encoding many memories, and each memory is represented across huge numbers of connections. Finally, retrieval follows a similarity-based principle that renowned memory psychologist Endel Tulving called the encoding specificity principle as far back as 1970.[1] The principle states that events are recalled to the extent the context of retrieval reinstates the context of encoding. If I say to you "money bank," wait a little while, and then ask what the second word I showed you earlier was and give you the hint "river," your likelihood of recalling "bank" is low even though "river" is more strongly associated with "bank" than "money" is. But if I say "savings," you will recall "bank" right away even though I never showed you "savings" before. "Savings" reinstates the semantic context of "money," making it a much more helpful cue at retrieval than "river." It is the similarity between encoding and retrieval contexts, especially their *semantic* similarity (how close they are in meaning), that determines whether you will be able to remember an event.

Like people, LLMs are driven by similarity. They are essentially extremely powerful memories that retrieve information in a way that depends on the retrieval cue, the prompt they are given to respond to. They are so sensitive to the retrieval cue that they give the appearance of not generating from memory but instead creating their output anew. This is true of people too. When jazz musicians improvise, the majority of what they produce are clips from pieces they have played or heard before ("ideas, licks, tricks, pet patterns, crips, clichés") that they are putting together in new ways.[2] When you participate in a conversation, most of what you produce are memories of things you have said earlier, witnessed, or heard about. There is nothing wrong with this. Constructing from memory can be a highly creative act and what you end up producing may in fact be novel. It takes a rich and sophisticated memory to make it happen.

My point is not that LLMs are the same as people. They wouldn't be artificial intelligences (AIs) if they were. The biggest difference is the size of their database. They have been exposed to orders of magnitude more information than people have and, correspondingly, have more storage capacity. And what they have been exposed to are not the events of a human life. It is surely impossible to think, or at least remember, like a human without having the very human experiences of parenting, death, and bad jokes in the specific order that is laid out by human development. LLMs and human beings differ in essential ways.

Here is another way they differ. Human beings have more than just sophisticated memories. They also have the ability to deliberate about, analyze, and reflect on what their memories generate. That is how we know that drinking beer doesn't make you young and beautiful, even though we have seen thousands of images in ads of young, beautiful people drinking beer. LLMs cannot reflect and deliberate in the same general-purpose way. They may not need to. They have such large stores of knowledge that they can often draw reasonable conclusions without as much deliberation as people (they are more likely to be able to retrieve a factoid about the negative health effects of beer). Moreover, they can be enhanced with routines that do careful deliberation in specific situations. Such routines may be in place already to prevent them from, say, instructing people how to build a dirty bomb. But AI has not made anything like the advances in the science of deliberation that they have made in the science of memory.

The chief way that LLMs are like people is that they make mistakes. LLMs are like a boosted version of everybody's Uncle Fred who always has something to say, on any topic, but half of it is wrong. LLMs are boosted in that they don't just have something to say, they have a lot to say on any topic, and perhaps less than half of it is wrong. After all, they have prodigious memories. But a lot of it is wrong.

What does all of this have to do with decision making? To make a decision, the decision maker has to consult their beliefs to figure out which outcomes could emerge from a choice option and how much confidence to have in each outcome. This entails a lot of cognitive work. How do people do it? Some human foibles and tendencies can be understood by understanding how LLMs work.

Predicting Outcomes

First of all, human decision making depends on memory. The most common strategy for decision making is habit, or choosing what you chose last time.[3] Doing so requires remembering what you chose last time. Remembering also enters the decision-making picture in many more subtle ways.

Consequentialist choice requires determining potential outcomes and judging their probability. The largest source of potential outcomes is memory. What outcomes have occurred in the past? What outcomes have others mentioned? When Arminius considered the potential outcomes of attacking the Romans, he must have thought about all the times the Romans had proven superior to their adversaries. He also must have considered battles involving surprise attacks and extended battle lines. Reasoning by analogy from memory is fundamental to human thought. When you consider the potential outcomes of going for a hike in the mountains, you retrieve from memory cases of people getting lost and freezing to death, and therefore you bring a good map and warm jacket.

Probability judgment similarly depends heavily on memory. There are several ways to judge the probability of an outcome:

- cold calculation
- rough calculation (e.g., Venn diagrams)
- judgmental heuristics
- mental simulation (causal and spatial models)

You can engage in cold calculation of a probability if you are betting on winning a raffle and you know the number of tickets that have been sold and the number you bought. Cold calculation is what scientists use to determine the probability of their empirical results. It is also what insurance companies use to determine the probability that they will have to pay out when selling policies. Generally speaking, cold calculation requires knowing a difficult mathematical theory, probability theory, and having the wherewithal and tools to implement it to do frequently complex calculations. You need the data to do the calculations on as well. There is nothing like cold calculation when you can use it. But rarely do we have everything required.

Rough calculation is an underused tool. Some of the famous fallacies of probability judgment could be prevented using rough calculation. Here is one of my favorite examples from the masters of the psychology of judgment and decision making, Amos Tversky and Danny Kahneman.

> Consider a regular six-sided die with four green faces and two red faces. The die will be rolled 20 times and the sequence of greens (G) and reds (R) will be recorded. You are asked to select one sequence, from a set of three, and you will win $25 if the sequence you choose appears on successive rolls of the die.

1. RGRRR
2. GRGRRR
3. GRRRRR[4]

Most people choose sequence 2 because it has the greatest ratio of green to red faces, and the die that is rolled has twice as many green as red faces. In that sense, sequence 2 is most *representative* of the outcomes of rolling such a die. But a little reflection shows that if sequence 2 occurs, then sequence 1 also has to occur (sequence 1 is nested within sequence 2; sequence 2 consists of G followed by sequence 1). So sequence 1 is more likely because it appears if sequence 2 does even if the initial G does not occur. This is known as a conjunction fallacy because sequence 2 is a conjunction of G and sequence 1. In probability theory, conjunctions can never be more probable than their constituents, so the fact that most people bet on sequence 2 is a violation of the conjunction rule of probability.

Note that representativeness is a common way of judging probability. After all, we use it to infer what objects are hundreds of times per day. If you walk into a room and see something with a seat, back, and four legs, it is probably a chair. If it looks like a chair and quacks like a chair, then . . . Judging by representativeness is just using our neural nets to classify not only objects but also abstractions like the outcomes of bets and other decisions. It relies on similarity and memory. A representative instance is a representation in memory that is similar to a prototypical instance of the category. All of this describes LLMs too.

Thinking from the Inside versus Outside

The incidence of this error can be reduced by thinking in a way that supports rough calculation. Instead of thinking about the similarity between each outcome and the process that generated it, you could think about all the instances that would involve each outcome. If you thought about all the sequences that included GRGRRR and then thought about all the sequences that included RGRRR, you would be more likely to realize that the latter set includes more sequences and is therefore more probable. Thinking in terms of similarity has been described as thinking from the inside because it involves thinking about the properties of an outcome (e.g., the number of Gs and Rs) and how those properties relate to the process generating the outcome (e.g., rolling a die with four Gs and two Rs). But thinking in terms of instances involves thinking from the outside in the sense that you are looking at the whole space of outcomes and each outcome is only one element of that space.[5]

Here is another example of the advantages of thinking from the outside rather than the inside. In 1988, German psychologist Klaus Fiedler gave a group of German students the following problem:

> Walter is 34 years old. He is intelligent, but unimaginative, compulsive, and generally lifeless. In school, he was strong in mathematics, but weak in social studies and humanities. Please rank order the following statements with respect to their probability:
>
> 1. Walter plays cards for a hobby
> 2. Walter is an accountant who plays cards for a hobby[6]

Like the green and red die example, there is a right answer here. The conjunction rule of probability dictates that the conjunction of playing cards and being an accountant (option 2) must be of lower probability than just playing cards (option 1) because if Walter is a card-playing accountant, then he also plays cards. Yet he might not be an accountant and still play cards. Most people committed the conjunction fallacy in this case by choosing option 2 because they judged based on representativeness. Judging from the paragraph, Walter seems like an accountant.

A small modification to the question led to very different results. When a different group of Fiedler's students were asked of the same two options, "To how many out of a hundred people who are like Walter do the statements apply?" many more people chose the correct answer (option 1). By posing the question in terms of a hundred people, Fiedler induced his students to think about the problem in terms of instances, from the outside. It was as if they had a little Venn diagram in their heads that looked like figure 5.1.

Judging from the description, Walter is clearly more likely to be an accountant than to play cards for a hobby and thus the area corresponding to the probability that he is an accountant is larger than the area corresponding to the probability that he plays cards for a hobby. Once the situation is thought about this way, it is obvious that the area of overlap between the two regions (where he is both an accountant and plays cards) is a subset of the card-playing area. Hence the probability of the conjunction is

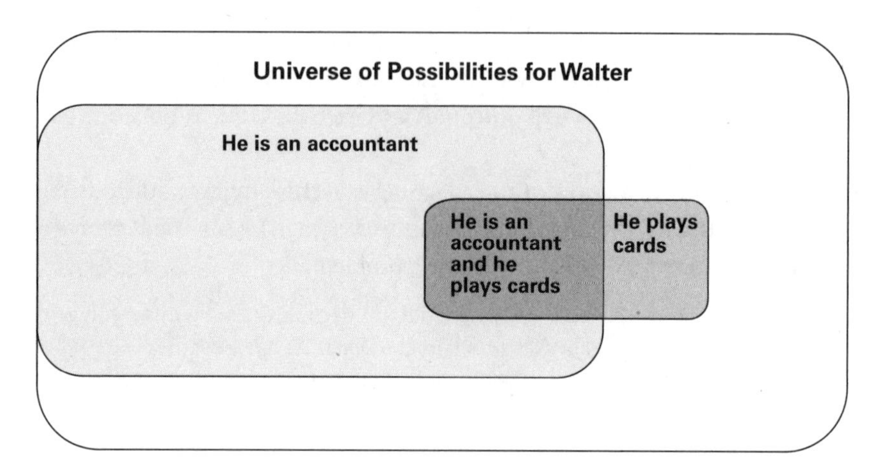

Figure 5.1

transparently less than the probability of the card-playing area. The conjunction fallacy is avoided due to a slight change in wording.

More on Memory and Probability Judgment

Memory comes into play in probability judgment in other ways. The most obvious is via the availability heuristic, the rule of thumb that we use to judge probability by evaluating the ease with which relevant instances come to mind. Everyone slows down on the highway when they come across an accident. First, they slow down because of the gawker's delay to see the demolition. But most drivers slow down for many miles as the accident remains available in their memory, reminding them of how dangerous highway driving can be. The availability heuristic also explains why people think that homicides are more likely than suicides. Homicides are publicized much more widely and therefore are more available in memory. In fact, suicides are far more common. In 2020 in the United States, a year renowned for its high murder rate, there were still more than twice as many suicides as homicides.

Another way we use memory to make numerical judgments is with the recognition heuristic.[7] Which city do you think has a larger population, Chongqing or Hong Kong? Most Westerners think the answer is Hong Kong because they have never heard of Chongqing or have heard very little about it. The fact that Hong Kong is recognizable inflates estimates of it. Yet 2022 estimates put the population of Chongqing at well over twice the size of Hong Kong. Memory is sophisticated, however, so if I ask you which is bigger, Chongqing or Charleston, South Carolina, you are likely to get it right (over 17,000,000 in Chongqing versus about 150,000 in Charleston in 2022). Memory delivers not only how recognizable a place is but also relevant facts. Here you probably recognize Charleston but you also retrieve the fact that cities in China tend to be large and that Charleston is not a huge city.

What is common to all of these cases is that people are using the same basic principles as LLMs to make the judgments. They are first querying memory and then spitting out an answer that does not necessarily conform to the dictates of probability theory. And human memory and LLMs store memories as well as query them in roughly the same way. In both cases, memories are stored as associations among their elements. Both humans and LLMs store the fact that wings go with feathers, steel goes with strength, and cleverness goes with success. They also store higher-order correlations, such as that the dependence of wings on feathers itself depends on having other bird properties as opposed to other mammalian properties (to exclude bats), and that the dependence of cleverness on success depends on having sufficient resources.

Other properties matter too. LLMs and people both have a kind of attentional mechanism, for example. But a lot of the commonality between LLMs and people is that both represent associations at multiple levels.

This dependence on associations explains why both systems are so quick to stereotype. Stereotypes are often assumed to capture central tendencies in a category. Sometimes they do. The stereotypical car in 2024 is not a Ferrari, nor a Tesla, nor a two-person Smart car. It is an average, boring four-door sedan with a gasoline engine. But many stereotypes are extreme instances of a category. The stereotypical diet food is celery even though the food people most commonly eat on a weight-loss diet might be rice or fruit. Still, celery is a good representative of diet foods because it is at an extreme on the key dimension for diets, calories. The stereotypical Hollywood actor is beautiful, rich, and famous. The average Hollywood actor may be beautiful, but they are neither rich nor famous. They most likely do some kind of menial work (perhaps serving tables at a diner on Sunset Boulevard). Our mental categories tend to capture salient properties and represent the categories in terms of our goals or the goals that we attribute to category members.[8] Instead of seeing averages, we see ideals.

An even more important function of our stereotypes is to let us know the properties that are shared by category members. The US stereotype of liberals is that they are wishy-washy, and believe that climate change is a major threat, people should have the freedom to choose whether or not to abort a fetus, citizens should be taxed to pay for welfare and infrastructure, and so on. Conservatives tend to be seen as governed by sacred values, as climate change deniers, and as believing that abortion should be illegal, taxes are rarely justified, and so forth. In both cases, what is important is that there are sets of properties that tend to go together (at least in observers' minds). These are the correlations that constitute mental representations in both humans and LLMs. They are also the correlations picked up by the implicit association test, known as the IAT, a test that purports to demonstrate implicit bias in attitudes but in fact has little or nothing to do with them.[9] What it picks up on are people's stereotypes. The fact that I have a stereotype does not imply any attitude toward it. I might have the positive stereotype that the New York Yankees are winners. But I still might hate the Yankees.

LLMs have a distinct advantage over humans when answering probability questions. They are more likely to have learned and remembered the correct answer. For instance, they are likely to be familiar with Chongqing and to know its population. They are also likely to have learned about the conjunction fallacy from the many discussions on the web about it and therefore might not be susceptible to it. Yet informal experiments with an older LLM (GPT-3) suggest that it makes some of the same errors that people do. The LLM ChatGPT in February 2023 showed the conjunction fallacy with both standard and new problems.[10] But early evidence suggests that newer versions of the LLMs do not make these errors.[11] They may be able to avoid them for the same reason that decision-making researchers and readers of this book avoid them, because they have learned the structure of the problem and the correct answer.

For the sake of completeness, I will point out that there are other heuristics that people use to judge probability and make numerical estimates. These include anchoring and (insufficient) adjustment as well as the proximity heuristic (we use distance to estimate risks and probabilities). I will leave it to the many other sources available on the heuristics and subsequent biases in judgment to elaborate on them.

Random or Reason?

Imagine you have decided to buy a house. To do so, you have to get some cash by selling some mutual funds—assets that you have tied up in the stock market. If you sell when the market is high, you could sell for far more than if you sell when the market is low. Therefore you have been keeping your eye on movements in the market so you can predict when it will be high and sell at the right time. One representative market index, the S&P 500, has had the following values for each of the past ten days, respectively:

4,289, 4,531, 4,603, 4,599, 4,470, 4,540, 4,621, 4,611, 4,638, 4,571

When should you sell? Should you wait? If so, how long should you wait for? Does it look like the market is increasing or decreasing? What is the trend? Maybe there is no trend. Perhaps these are just random fluctuations. If so, it does not matter when you sell. All you can do is roll the dice.

This decision involves, as a first step, deciding whether the pattern we see is random or not. What does it mean to say that a pattern is random? Many computer languages offer a way of generating a random number, often called a random seed. One way they do so is to pick the next number in a sequence from a table of random numbers. But there is no way to know what makes the numbers in the table random. Another way is to pick a number that depends on some random event. For example, you might generate a number based on the current time assuming that you do not know what time it will be when the program is run. In both cases, "randomness" is really being defined in terms of absence, the absence of any ability to predict what the outcome will be. In the case of a random seed, the trick is to pick a number that you cannot predict even though you know the causal process generating it.

Randomness really has to do with the knowledge of the person who is making a judgment or decision. If the person cannot say what the outcome will be, then it is random to that person. So if I choose a number between 1 and 100, and ask you to guess what it is, then that number is not random to me (I know perfectly well what it is) but it might be random to you—though it is likely not perfectly random. It is unlikely to be 1 or 100, any multiple of 10, or any number with the same digit repeated (11, 22, 33, etc.). People tend to avoid such numbers when asked to generate a random number. That actually rules out a lot of numbers, so you should do much better than chance. Chance implies that you could only guess it 1 out of 100 times, but I have just reduced the set of possibilities from 100 to 80.

When making a decision, people rule out randomness when they believe there is structure they can deploy. Ruling out randomness when looking at stock market data over time means that you believe you can see a trend. It has been well established that people see trends that do not exist. A classic example was presented by psychologist Tom Gilovich and his colleagues in 1985. They asked their subjects the following question:

Which of the following sequences of X's and O's seems more like it was generated by a random process with Pr(X) = Pr(O) = .5 (e.g., flipping a coin)?

a. XOXXXOOOOXOXXOOOXXXOX

b. XOXOXOOOXXOXOXOOXXXOX[12]

Most people thought that sequence b looks more random. Both of these sequences are just sequences of *X*s and *O*s written down by a human being, so neither is truly random. Nevertheless, one property a truly random sequence should have is that events are unpredictable. Both of the sequences have 11 *X*s and 10 *O*s, so the probability of the letters does not help. One insight that might help is that if you know the sequence up to a certain point, then you should not be able to predict the next element. In the sequences shown, this would be true if the probability of one element, say an *X*, was the same regardless of whether the preceding element was an *X* or *O*. In a truly random sequence, whether an *X* or *O* comes up, the following letter has the same probability of being an *X*. You can verify for yourself that in sequence a, the probability of an *X* following an *X* is 5/10 and the probability of an *X* following an *O* is also 5/10. So the criterion of unpredictability is satisfied by sequence a. For sequence b, the probability of an *X* following an *X* is 3/10 and the probability of an *X* following an *O* is 7/10. *X*s are more likely following *O*s than following *X*s. So sequence b is not completely unpredictable. If you see an *O*, you should bet that the next symbol will be an *X*. In that sense, sequence b alternates more than a purely random sequence would on average.

Yet people see sequence b as more random than sequence a. Why? Gilovich and colleagues think it is because sequence a has more streaks (a streak of 4 *O*s and two streaks of 3 *X*s) than sequence b (one streak of 3 *O*s and one of 3 *X*s). Streaks signal nonrandomness. When an event happens multiple times in a row, we jump to the conclusion that there is a reason for it. For instance, when a team wins multiple times in a row, they seem to be on a roll—a winning streak—and not winning by chance, even though if they were winning by chance, they would still win multiple times in a row on occasion. The streaks in sequence a make people think they were generated by some causal process and not by something outside the realm of explanation, randomness, even though random sequences are guaranteed to have some streaks. We see structure in sequences that is not there whether we are talking about sports events, the stock market, or the climate.

More on Seeing Structure Where It's Not

Imagine you are playing roulette at a casino. The wheel has come up black six times in a row and you are deciding whether or not to bet on red. It seems like red is due. After all, roulette wheels offer the same chance of black and red. Black has come up several times in a row, and it is time for a red. Maybe you should put all of your chips on red?

Maybe not. I don't believe in the supernatural, so I don't think there's a mystical force determining whether roulette wheels show black or red. Moreover, I do not believe that casinos rig the game because they don't need to. They make plenty of money without doing so, and if they did rig it and got caught, the penalties would be severe. So I think the outcomes of roulette wheels are random—that is, unpredictable. Streaks of six happen on occasion. In fact, if you spin the wheel enough times, a streak of six is guaranteed. So the likelihoods are exactly the same of the wheel coming up black or red following a streak of six.

Most people are not governed by the logic of randomness. They believe that if six blacks in a row come up, then red is due, and they bet on red. This belief, that chance is self-correcting, that random sequences right themselves, is called the gambler's fallacy. It shows up all over the place. It has been shown in horse racing, blackjack, playing the lottery, and online gambling.[13] The fact is that random events have no memory so they cannot self-correct. The gambler's fallacy arises because people believe that random sequences appear random in small, local windows of time, not just globally, where a large number of events guarantees events will be distributed randomly.

Here's a final example of randomness. You go to a friend's house for dinner. You discover name tags at each dinner setting. You have been placed beside an attractive person, the host's sibling, who you have been eyeing for a while—and the host knows it. But you don't know if this other person is attached to someone else or not. Did the host intentionally seat you next to this person? Or was it a random coincidence? Knowing the host's intention could make a difference to your behavior over dinner.

How many of us would assume that the seating was random? The tendency to ascribe hidden motives to the host is powerful. After all, the probability of a coincidence is low. Of course, the probability of some coincidences happening over the course of a lifetime is high. The point is that we see structure in random events whether those events occur over time as in the S&P 500 and roulette wheel examples, or laid out in space, as in the dinner seating example.

A Vicious Cycle

We have now seen multiple examples of how the design of human cognition leads to beliefs that bias judgment away from normatively justifiable responding. Do they contribute to conflict and polarization, one of the themes of this book? They do, in part, by making people less cognitively flexible. People tend to be intransigent. In the face of new data, we tend to stick to our guns. This is often referred to as "confirmation bias," but confirmation bias is really a cluster of related biases.[14] One bias is that when we are shown data that do not support a hypothesis that we currently hold, we don't change our minds. Sometimes we even become more sure of what we already believed. The point was made elegantly by an experiment reported in 1977.[15] The experimenters showed subjects two urns. One urn contained seventy red balls, thirty white balls, and fifty blue balls. The other urn contained thirty red balls, seventy white balls, and fifty blue balls. The experimenters then put the urns behind a curtain and started taking pairs of balls from one of them, showing the sampled balls to the subjects. The subjects' task was to guess which urn the balls were coming from. In some cases, the subjects would have seen a lot more red than white balls and so would be pretty sure that the balls were coming from the first urn; they judged the probability of the first urn to be high. Then the experimenters would sample a red and white pair and ask the subjects what the probability was that the balls were coming from the first urn. Should a red and white sample change anyone's mind? The first urn has seven reds for every three whites, and the second has the opposite, seven whites for every three reds. So the probability of both a red and white has to be identical for the two urns. A red and white pair is not diagnostic; it doesn't help the subject at all. Similarly, if the experimenter showed two blue balls, that would provide no information because both urns have the same number of blue balls (fifty). But in both cases, showing either the red and white sample or the two blue balls increased subjects' rating of the probability that the balls were coming from the first urn. Even though the data they were shown in no way supported their hypothesis, they treated it as evidence in favor of the hypothesis they were already holding. Why? Well, if the balls were coming from the first urn, they would expect a red and white on occasion. Similarly, they would expect two blue balls on occasion (after all, there are fifty of them). So the samples satisfied subjects' expectations regardless of the information in the samples, leaving subjects feeling more sure of themselves.

This tendency to interpret evidence in a way consistent with what we already believe has also been shown with political issues, first with capital punishment and more recently with affirmative action and gun control.[16] Studies demonstrate that people find evidence more credible if it supports attitudes they already have rather than provide evidence that they are wrong. Therefore supporting arguments have more influence on the beliefs that they espouse at the end of the study.

Not only do people interpret evidence in a way that supports what they already believe, they tend to choose evidence that is consistent with their view. When Dick Cheney was the Republican vice president of the United States, he insisted that the hotel rooms he was staying in had their TVs tuned to the Republican-leaning Fox News before he arrived.[17] This kind of selective exposure to content has been demonstrated experimentally. For instance, a study published in 1961 asked mothers to choose to listen to a speech advocating that genes were the primary driver of child development or one arguing that environmental factors were. The women chose whichever speech was most consistent with the position they already held on the issue.[18]

A wide-ranging review of studies of selective exposure concluded that people have a moderate preference for new information they agree with over new information they do not agree with.[19] Presumably the reason people have only a moderate preference is that people do appreciate the value of learning. We frequently learn more when we hear opposing positions as they tend to provide more unfamiliar perspectives. This desire to learn works against our desire to have our opinions confirmed. Despite the value of opposing viewpoints, modern social media exposes us far more often to views we agree with than those we disagree with, in part because people choose outlets that are aligned with their attitudes and in part because the algorithms that select what to show on a person's feed tend to favor what the user has liked in the past. This is one reason that we tend to reside in echo chambers, interacting only with those who feed back to us what we already believe.[20]

When we cannot find any other way to get biased evidence that supports what we already believe, we turn to our memories. We have a habit of recalling events that are consistent with our beliefs.[21] One experiment asked people to remember occasions in their lives when they had been extroverted or introverted. But first, half of the people were shown evidence suggesting that extroverted people are more successful than introverts. That group remembered more examples of their own extroverted behavior than introverted behavior and recalled the extroverted examples more quickly. The other half of the participants were first convinced that introversion led to greater success than extroversion and showed the opposite pattern of recall.[22]

In sum, people's cognitive systems are designed to support the beliefs their owners already have, even in the face of contradictory evidence or arguments. When we surround ourselves with those of like mind, individuals can feed on each other, finding mutual support for positions regardless of whether they are right or wrong, thereby feeding each other's confidence. If the bubble we live in becomes sufficiently large, homogeneous, and rewarding, we can lose sight of even the existence of opposing perspectives and become dismissive of them when we do confront them. This is how vacuous ideologies of the type often seen in totalitarian states form. It also explains the emergence of conspiracy theories everywhere. The cognitive system reasons more effectively when it is not in an intellectual bubble but instead evaluating beliefs in a test bed of diverse perspectives that offer conflicting evidence and arguments. Operating in accompaniment with this kind of intellectual diversity can negate our confirmation biases.[23] Sometimes, though, we don't have the luxury of surrounding ourselves with a range of opinions. When we have to reason alone, it turns out that testing our ideas by trying to explain to ourselves how our beliefs lead to consequences can reveal to us what we don't know.[24]

AI, Memory, and Belief

LLMs were not always the cutting edge of AI. For most of the history of AI, people put their money not on incredibly sophisticated memories like LLMs but instead on systems that could reason. Over the course of time, AI researchers tried fancier and fancier kinds of logic to build intelligent systems. My own bet twenty years ago was that causal logic would do the trick.[25] Doing any kind of logic involves making explicit assumptions about the world and deriving clear conclusions from those assumptions. The whole path of derivation can be traced so that what the system is doing is transparent. Its reasoning can be observed, reproduced, and reported. Nothing is mysterious.

But LLMs are deeply mysterious. Nobody fully understands how they work. Their capabilities surprise even their creators. They are not doing the kind of logic that logicians do. They are encoding, storing, and retrieving information from memory and putting it together in ways that just happen to allow them to perform amazingly well amazingly often and to sometimes make ridiculous errors. Just like people.

We have seen in this chapter that people also rely heavily on memory when making decisions. We have also seen that people make a variety of systematic errors in that they violate the prescriptions of probability theory and good scientific practice. This suggests that, like LLMs, people rely not on logic but rather on their storehouse of knowledge, along with tricks that allow them to turn their knowledge into good guesses about the way the world is.

But humans are not limited to memory. We can learn logic and learn to deploy it to make better decisions. For instance, we can learn probability theory and use it to avoid some cognitive illusions. We seem to be better at that than LLMs. At least, that is true at the moment, as I am writing. Is it still true at the current moment, as you are reading?

6 Beliefs About How Things Work

Much of the conversation about decision making concerns games of chance—a move this book is itself guilty of. One reason for this reliance on gambling is, first, that the study of decision making has been heavily biased in favor of consequentialism and gambling is all about consequences, winning and losing. A second reason is that games of chance are simple test cases for thinking about decision making. Both uncertainty and outcomes manifest in the simplest possible way in a game of chance. The uncertainty—the probabilities involved in a game—are sometimes easy to figure out, and sometimes even announced by the state or company offering the game. And when it comes to outcomes, the value of money is a particularly simple way to think about value because the more money you get, the greater its value. People just want to win as much money as possible. Furthermore, the utility of money generally has an easy-to-understand shape, the smooth concave function shown in chapter 4.

But gambling fails to capture the heart and soul of the decision-making process. Rarely do people outside casinos choose with clear probabilities and regular, smooth utility functions. Most people do not even know how to calculate a probability, and most tough decisions are not solely about winning money. When people make a decision based on consequences, they are more likely trying to solve a complex problem about both the effects of the various options before them and what they really want out of the decision. This chapter will do a deep dive into how people determine consequences. One implication will be that a substantial portion of our judgments and decisions do involve consequentialist thinking; we are not governed exclusively by sacred values. The next chapter will focus on what people want.

How to Understand Causality

The consequences of most decisions are opaque. What will happen if you give your boss an ultimatum to give you a raise or you will quit? If you patch up the crack in your blender with duct tape, will you solve a problem or create a much bigger one? How should the United States respond to China's provocative actions around Taiwan? In each case, the question is a causal one. What will be the causal effect of taking an action as opposed to taking a different action or not acting at all? A big part of how people make a decision involves causal reasoning, reasoning about how causes lead to effects and what effects tell us about causes. In the case of decision making, the causal question is most often about how actions lead to consequences.

The currency of causal reasoning is causal beliefs. Why would your boss give you a raise? Because of your boss's causal belief that your presence in the company is critical to its performance. Why will duct tape prevent a crack in your blender from opening up? Because of your causal belief that duct tape has the strength to hold plastic together even under agitation. Why would US saber-rattling if China threatens Taiwan with aggression have negative repercussions? Some pundits have the causal belief that any show of force will unleash a fury of Chinese patriotism leading to violence.

Cognitive scientists define causal beliefs in different ways, but broadly, a causal belief is an answer to a why question. This definition is not sufficient, however. Some answers to why questions do not seem, on their face, to concern causality. One answer to the question "Why would your boss give you a raise?" might be that they are a generous person. This answer is not really causal. It may have some causal implications, but the answer is attributing a property (being generous) to a person (your boss). That does not seem to be a causal statement. A second problem with this definition of causality is that it begs the question. What makes the belief that "your presence in the company is critical to its performance" causal? Similarly for the beliefs about duct tape and about China. I simply asserted they are causal without justification.

Three theories of causality provide more substantive answers. One theory states that causal claims depend on the truth of a particular counterfactual, a statement that is not true in the world currently being considered but that would be true in an alternative world.[1] To say that A causes B is to say that A occurs and so does B, and if A had not occurred, then

B would not have either. In the alternative world without A, B would be absent too. Your boss's belief that "your presence in the company is critical to its performance" is a causal belief because it implies that you are present and the company is performing well. In addition, there is a counterfactual belief that if you were not present, the company would not perform well. To avoid that alternative world, your boss should keep you. The belief that "duct tape has the strength to hold plastic together even under agitation" is causal because it implies that in the focal world, duct tape will be present and the blender will remain whole, and in the alternative world without duct tape, the blender would crack and send its contents all over the kitchen. The belief that "any show of force will unleash a fury of Chinese patriotism leading to violence" is causal because of the implied counterfactual that the alternative world without a show of force will not have violence.

This counterfactual view of causality makes sense, but it is hard to distinguish from claims about mere correlation. If A is strongly correlated with B, then it also seems to be the case that worlds with A will have B and worlds without A will tend not to have B. The main goal of theories of causality is to distinguish causation from correlation. When making decisions, causation provides a reason for choosing but correlation does not. If I choose to take a French course because I want to learn French, then the course better cause me to learn to speak French. If it is merely correlated with speaking French because people who take the course happen to already speak French, then I'll be wasting my time.

To remedy this problem, some theorists argue that causal statements do not just imply a counterfactual; they also imply consequences of an intervention.[2] On the interventional view, to say A causes B is to say that if an agent intervenes to make A occur, then B will occur too. And if the intervention prevents A, then it will prevent B as well (unless B occurs for some other reason). On this view, your boss's belief that "your presence in the company is critical to its performance" is a causal belief because it implies that any intervention that forced you to be present in the company would imply a world with a well-performing company, as long as the intervention did not itself hurt the company. Furthermore, it implies that an intervention leading to your absence would imply a world with the company performing poorly. The belief that "duct tape has the strength to hold plastic together even under agitation" is causal because an intervention that applies duct tape will also hold the plastic together and one that prevents the application of duct tape will put us in a world where the plastic falls apart. A parallel logic applies to the China example.

A complaint about this account of causation is that, like the first account, it is circular. After all, what is an intervention? It is hard to answer that question without referring to causation. I must admit that, despite the circularity, I am partial to the interventionist account because it grounds causality in the actions of an external agent and that strikes me as exactly how causality should be grounded. The best way to infer causation in science—and life—is by using the experimental method. The experimental method grounds the inference to a cause in intervention. An external agent in the form of an experimenter manipulates—intervenes on—an independent variable. Subsequent effects on dependent variables can be causally ascribed to the experimental manipulation.

Despite my preferences, philosophers do not like circular explanations and therefore some of them do not subscribe to the interventionist account. They might instead rely on a third idea about causality, that A causes B if there is some conserved quantity—like force—that travels from A to B over time and space.[3] This is sometimes called the power theory of causation. It changes the justification for calling statements causal. The claim that "your presence in the company is critical to its performance" still qualifies as causal on the assumption that there is some quantity—your effort or good nature perhaps—that travels from you to others in a way that generates good performance. "Duct tape has the strength to hold plastic together even under agitation" is a causal statement because it asserts that duct tape creates a force that holds the blender together. The force is not really traveling, it's true, but it is transferred from the duct tape to the plastic blender to influence the physical state of the blender. The last case, the belief that "any show of force will unleash a fury of Chinese patriotism leading to violence," is causal on the conserved quantity view because symbols can serve as conserved quantities. It is the symbolic content of a show of force that gets transformed into Chinese fury.

Are these three interpretations of causality—the counterfactual, interventionist, and power theories—complementary or mutually exclusive? If one is true, does that mean the others are also true or that they are false? The interventionist theory implies a counterfactual one, in that if intervening on A causes B, then the counterfactual "if A is not intervened on, then B would not occur" should be true. This suggests those two at least are complementary. The power theory implies a counterfactual too. Namely, it implies that if a conserved quantity were not passed from cause to effect, then the effect would not occur. So those two theories seem complementary as well. But the interventionist and power theories offer divergent explanations for what makes a relation causal. The interventionist theory appeals to what would happen under an agent's control, and the power theory appeals to the precise mechanism that the cause uses to drive the effect.

Psychologically, the two may apply in distinct circumstances. Studies suggest that people tend to rely on power theories when reasoning about physical situations but not when reasoning about effects that result from human intention.[4] If a boulder rolls down a hill and crushes a car, then the boulder is causally responsible because its rolling transferred the force that crushed the car. The power theory provides the best explanation. But if I pushed the boulder and set it off, I am causally responsible. This is not because I exerted a force powerful enough to crush the car (I am not that strong). It is because of a counterfactual that follows from my intervention. If I had not pushed the boulder, the car would not be crushed.

There is also reason to believe that people reason using the logic of intervention when making decisions.[5] Imagine you want to improve your health and you think eating broccoli may be good for you. If you learn that, in fact, eating broccoli does improve your health, then you should choose to eat more broccoli. But what if all you know is that there is a correlation between eating broccoli and being healthy? In that case, eating broccoli might not cause good health. Perhaps people who eat a lot of broccoli are just more health conscious, and tend to eat well generally and engage in other healthy behaviors. In that case, eating broccoli is not a cause of good health but rather a third variable, being health conscious, is a joint cause of eating broccoli and of being in good health. That set of beliefs provides much less reason to take up a broccoli habit if your goal is to improve your health. Because you have chosen, as an independent agent, to change your diet, eating broccoli will no longer indicate anything about how conscious you are about your health and therefore will have nothing to do with how healthy you are.[6] It turns out that people are highly sensitive to the difference between these two sets of causal beliefs. They are much less willing to start eating broccoli if it is only an effect of a third variable and not a cause of the goal they seek to satisfy. They want their intervention (eating broccoli) to have the desired effect.

The Critical Role of Causality in Interpreting Evidence

Now that we have an idea of what we mean by causation, we can address some of the more subtle benefits that causal thinking has. One is that causal beliefs turn out to be necessary to make reasonable scientific inferences. This case has been made in a compelling way by the leader in the modern study of causality, computer scientist Judea Pearl. His argument concerns an apparent paradox of statistics called Simpson's paradox.[7]

Let's say we're doing a study of whether a drug is effective. We give the drug to a group of 40 sick people and 50 percent of them recover. We need a control group to make sure that people are not just recovering even without the drug, so we give a placebo to an additional 40 sick people. Only 40 percent of them recover. Is the drug effective? Perhaps. The difference between the groups is not large, but it favors those who got the drug. Now we look at the data more carefully. All the people in the study were either men or women. We find that 60 percent of the men who got the drug recovered, but 70 percent of the men who got the placebo recovered. Of the women, 20 percent who got the drug recovered, but 30 percent of those who got the placebo did.

First question: Is this possible? How can it be that a greater proportion of the group that got the drug than the group that did not recovered, yet when we divide the groups up into mutually exclusive and exhaustive subsets, we find the opposite in both groups? In both subsets (men and women), a greater proportion who did not get the drug recovered. The answer to this question is simple: Yes, it is possible. Here are the number of people who recovered over the total number of people tested in each group:

	Males	Females
Drug	18/30	2/10
No drug	7/10	9/30

You can see for yourself that 20/40 who got the drug recovered (50 percent), 16/40 who did not get the drug recovered (40 percent), and the proportions of both males and females who recovered is greater for those who did not get the drug than for those who did get it. There is nothing actually paradoxical here. The strange findings are due to the fact that men recover at a higher rate than women (25/40 versus 11/40), and more men were put in the drug condition and more women were given the placebo.

Here is the second question: Is the drug effective? Now that we see the breakdown by men and women, I would say absolutely not. For everyone who is either a man or woman, the drug is not only ineffective but makes recovery less likely too. It only seemed to be effective before we compared men and women because of a bad experimental design—the drug and no drug groups had different numbers of men and women.

Compare this conclusion to a similar study with equal numbers of men and women in the two groups. In this study, we also measured people's blood pressure after they had taken the drug. Imagine we compare two groups again, but this time we compare those with low blood pressure and those with high blood pressure, getting exactly the same results as we did earlier comparing men and women:

	Low blood pressure	High blood pressure
Drug	18/30	2/10
No drug	7/10	9/30

Again we have the odd result that more people recover with the drug than without, but the opposite is true for both of the two subsets (those with low and high blood pressure). For this second study, we again ask whether or not the drug is effective. My answer is different this time. I say it is effective and likely cures people by lowering their blood pressure. Of those who got the drug, 30/40 have low blood pressure. Of those who got the placebo, only 10/40 have low blood pressure. Because we took their blood pressure after administering the drug or placebo, a reasonable conclusion is that the drug had the causal effect of lowering their blood pressure. And more people who had low pressure recovered than those with high pressure, suggesting that lowering blood pressure is an effective way to cure the illness.

Why did I not draw the same conclusion when we were comparing men to women? Because in real life, drugs (alone) cannot change people's sex.

There is no causal link from taking the drug to whether the patient was a man or woman so that mechanism for the drug's action is ruled out. Drugs can reduce blood pressure but they cannot change sex. These are my causal beliefs, and they are critical for interpreting the study and deciding whether or not the drug is effective at curing the disease. We cannot just dismiss causal beliefs and rely on statistics when drawing conclusions from evidence. Indeed, we are obligated to use causal beliefs to reach valid conclusions.

Everything Is Causal

So far, we have seen that causality can be grounded in three different ways (as a counterfactual, intervention, or conserved quantity), that people are pretty good causal reasoners, and that—in certain cases—the only way to interpret data correctly is to use causal knowledge. Now I will try to convince you that causality is a primary concern of people when they are making judgments and decisions. Let's start with an old demonstration from Amos Tversky and Danny Kahneman.[8] Imagine there's a mother-daughter convention with hundreds of mothers, each with one biological daughter. We sample one mother-daughter pair at random and look at the color of their eyes. Which of the following is more likely: (a) that the daughter has blue eyes if the mother has blue eyes, or (b) that the mother has blue eyes if the daughter does. This is a slightly tricky question. Most people think the answer is a because they imagine that blue eyes are transferred from the mother to the daughter genetically—a kind of causal process. To explain b, the mother having blue eyes if the daughter does, we have to think not in a causal direction, from cause to effect, but rather diagnostically, from effect to cause. That is harder. Thinking in the causal direction is natural and automatic.

As long as the proportion of mothers with blue eyes is the same as the proportion of daughters with blue eyes, however—surely it is if they are each a representative sample—choosing option a is a mistake. Both statements are equally probable. If the probabilities of a and b are the same, then the conditional probabilities of $P(a|b)$ and $P(b|a)$ are also the same. But that is not how people judge them. People are using causal strength, the ease of imagining transferring the blue eyes gene from one woman to another, as a substitute for judgments of probability.

So along with the other heuristics we discussed in the last chapter, people employ a causality heuristic to judge probability. They also employ a closely related heuristic, the simulation heuristic.[9] The natural way to answer questions about how familiar objects in the physical world would interact is to run a mental simulation. If I ask you whether you could squeeze twenty college students dressed as clowns into your car, a good way to come up with an answer is to imagine twenty clowns draped over one another filling the space inside your car. Can you squash twenty mental clowns into your mental car? Golf professionals use mental simulation to choose their stroke when putting. They look at the lay of the land and mentally simulate the ball's trajectory. Simulations are an important tool whether they are mental or not. For instance, computer simulations allow meteorologists to generate weather forecasts.

Here is my favorite simulation example from the Kahneman and Tversky oeuvre:

> Mr. Crane and Mr. Tees were scheduled to leave the airport on different flights, at the same time. They traveled from town in the same limousine, were caught in a traffic jam, and arrived at the airport 30 minutes after the scheduled departure time of their flights.
>
> Mr. Crane is told that his flight left on time.
>
> Mr. Tees is told that his flight was delayed, and just left five minutes ago.
>
> Who is more upset?[10]

The response to this question is almost universal. Mr. Tees is more upset. Despite the fact that the two men started in the same place with the same intention and ended in exactly the same position (stranded at the airport), Mr. Tees can more easily simulate a desirable counterfactual world—one with him flying off. The ease of generating a counterfactual serves as a powerful cue for the mens' emotional states.

Such counterfactual simulations can be useful learning tools. If Trump hadn't spoken to the crowd on January 6, 2020, would the crowd have attacked the Capitol that day? Answering this question has a couple of effects. First, it forces you to develop your causal model of what drove the crowd on that day. Simulations are generated by mental causal models, sets of beliefs about how the world works. If your causal model is too poorly developed to run a simulation, then you have to fill in the details, either by making them up or by doing some research. The second effect of this question is to reveal your predispositions. Trump haters are likely to answer "no, Trump was responsible for the attack," and may not even bother running a mental simulation. Trump lovers might actually deny the premise of the question, claiming there was no attack that day. They might not run a mental simulation either.

The examples so far all concern judgment, the fact that people use causality and simulation heuristics to determine how much certainty to have in events. Causal considerations also govern much of our decision making in a direct way. When faced with familiar decision problems, people often know what to do immediately and intuitively because, like modern AI, people are terrific pattern recognizers. This has been shown by psychologist Gary Klein in cases where decision makers are experts operating under extreme time pressure to save lives and property. He studied how firefighters make decisions. Here is one of his stories:

> [A] firefighter led his men into a burning house, round back to the apparent seat of the fire in the rear of the house, and directed a stream of water on it. The water did not have the expected effect, so he backed off and then hit it again. At the same time, he began to notice that it was getting intensely hot and very quiet. He stated that he had no idea what was going on, but he suddenly ordered his crew to evacuate the house. Within a minute after they evacuated, the floor collapsed. It turned out that the fire had been in the basement. He had never expected this. This was why his stream of water was ineffective, and it was why the house could become hot and quiet at the same time. He attributed his decision to a "sixth sense."[11]

Pattern recognition is the basis for quick, intuitive decision making, especially when decisions are familiar or one has lots of expertise. Intuitions have a lot of causal information that experts are able to access quickly, sometimes without even realizing that they are doing so. But sometimes intuitions do not do the job. In situations that are novel and unfamiliar, we have to more thoughtfully deliberate to develop our causal analyses. People who have spent hundreds of hours under car hoods may have no trouble figuring out why a car won't start, but the rest of us have to think carefully or reach out for help.

Causal Illusions

People are so enamored of causal thinking that we assume there are causal relations that do not in reality exist. Ellen Langer is a famous psychology professor at Harvard, known these days for her work on mindfulness. Earlier in her career, she showed how people experience an illusion of control, thinking that they had control over outcomes that they had no influence over, as long as there were cues present to make them think their skill was relevant. In one study, she found that when rolling dice in craps, people tend to throw harder for high numbers and softer for low ones. Clearly, how hard you throw dice has no influence over the outcome. In another study, she had people guess a series of coin tosses. When guessing, some people are going to be more successful than others just by chance. What Langer found was that those who were more successful began to believe that they were actually better guessers and that their guessing performance would be less accurate if they were distracted. They attributed their performance to their own skill rather than to what actually generated it, luck. She referred to these findings as the illusion of control.[12]

The illusion of control is the tendency to ascribe to ourselves causal powers that we do not have. We see it in operation in calls for war. Hawks overestimate their control over the outcome of a battle, and assume a quick and easy victory.[13] Hence hawkish US military analysts predicted that the Iraq War would be a "cakewalk" before they discovered it was anything but.[14]

The chief of staff for the French Army at the outset of World War I, General Noel de Castelnau, declared, "Give me 700,000 men and I will conquer Europe."[15] In fact, the French Army called up nearly 9,000,000 soldiers and did not conquer Europe.

The search for causes goes beyond the illusion of control. Kahneman argues that people spontaneously search for causes whenever they are confronted with an unexpected, threatening, or norm-violating event.[16] Superstitions provide an example. They are causal beliefs. Superstitions are ways of attempting to control or predict the future. For most of us, one or two superstitions are hard to resist, whether it is knocking on wood to prevent a bad omen or wearing a specific old cap when your favorite sports team is playing. In a poll, only 31 percent of people in the United States said they have no superstitions.[17] Superstitions are illusory in the sense that they are false causal beliefs, but unusual illusions in the sense that we generally know they are false yet act on them anyway.[18]

Other kinds of causal illusions are more subtle. Some even affect our perception. This has been known since the early part of the twentieth century. In a classic work, Belgian psychologist Albert Michotte used a primitive kind of animation to construct cartoons of blocks moving on a screen.

When one block made (virtual) contact with a second block and the second block then moved off, people reported seeing a causal event, the first block knocking the second block forward.[19] This perception of causation is one basis of our ability to watch videos and movies. We know we are watching pixels on a screen turn on and off, yet we perceive dramas, reality TV, and football games. We interpret changes in light as causal events.

Causal beliefs also influence us at a conceptual level. The classic demonstration of this was reported in 1969 by Loren and Jean Chapman, a married pair of clinical researchers.[20] They were working during a period in which homosexuality was viewed by many as a clinical syndrome. At the time, people were generally not vocal about their homosexuality; often, they were not aware of or would not admit their own homosexuality to themselves. Clinicians would frequently "diagnose" a client's homosexuality using a projective test. A projective test is a psychological test that is assumed to offer a means for clients to project important but latent facts about themselves onto their responses. A well-known example of a projective test is the Rorschach inkblot, developed by Swiss psychiatrist and psychoanalyst Hermann Rorschach around the turn of the twentieth century. The test consists of a standard set of inkblots—images formed by dropping ink on a page and folding it—and asking clients what they see. Then the analyst interprets the client's responses. The assumption is that an individual will classify the inkblots based on their emotional needs, base motives, and conflicts, the deep underground foundation of their individual psychology.

Projective tests have useful purposes, but in the 1960s they were frequently overinterpreted. Indeed, some responses to Rorschach inkblots led some excellent clinicians astray. For instance, clinicians persisted in reporting certain projective test diagnoses despite strong evidence that the diagnoses were invalid. The Chapmans wanted to know why and took the diagnosis of homosexuality with the Rorschach as a test case. Responses on the test actually did include some valid ones. The coauthors referred to them as type E signs. These were not face valid but rather empirically valid. For example, for unknown reasons, homosexuals tended to be more likely than nonhomosexuals to see monsters on card IV, a specific Rorschach inkblot. They were also more likely to see a figure that appears to be "part animal" and "part human" on card V. But these were not the cards that clinicians relied on to classify their clients. Instead, they used type F signs,

signs that were face valid but empirically invalid. They tended, for instance, to classify people as homosexual if they reported anal content, genitalia, or feminine clothing on any of the cards. Apparently, clinicians did not generate diagnoses on the basis of scientific evidence in the form of valid correlations between what people see in the cards and their actual diagnoses. They instead used illusory correlations—expectations based on semantic associations or causal beliefs that were not actually present in the data.

To show this systematically, the Chapmans first wrote to clinicians who used the Rorschach test and invited them to fill out an anonymous questionnaire asking them to list examples of the kinds of responses they had obtained from men with homosexual impulses. The five most common responses reported back were all type F. Only two of thirty-two clinicians ever listed a type E. This is evidence that professional users of the Rorschach test only saw responses consistent with their causal expectations, rarely reporting responses that had supporting evidence. The researchers also asked laypeople to rate the tendency of homosexuality to "call to mind" each of the signs. Type F responses were rated as moderately strong but type E as weak. So the professional clinicians had exactly the same associations as laypeople despite, in several cases, many years of practice.

In another study, the researchers fabricated materials consisting of thirty Rorschach cards, along with a response to each card and diagnosis of the client giving the response. These fabricated materials included the true correlations: Type E signs were associated with homosexuality and type F signs were not. Naive participants saw the cards and then were asked to indicate which symptoms were associated with homosexual descriptors. What they found was that the fabricated materials elicited the same descriptors from laypeople that the clinicians had mentioned. Both experts and laypeople saw what they expected based on their causal beliefs, not what was actually present.

This kind of illusory correlation is pervasive. Is there really a link between a full moon and how people behave? There is no evidence for one.[21] Measles, mumps, and rubella vaccines are not correlated with autism.[22] Republican presidents are not associated with a strong economy.[23] Diversity training is not associated with decreased prejudice.[24] In each case, opinion is guided by theory and expectation, not by data.

We have now seen a variety of examples of people giving priority to causal beliefs over statistics even when the statistics have more scientific evidential value. Do you make that error? Here are some statistics from the National Highway Traffic Safety Administration: About 37 people in the United States die every day in drunk-driving crashes, or 1 person every 39 minutes. In 2021, 13,384 people died as a result of drunk driving.[25] Imagine that you are at a friend's house and are offered another drink. You have had one or two already, but feel fine to drive right now; you do not think you would cause an accident. But you also know these statistics. Do you take the drink and drive home, take the drink and call a costly Uber, or refuse the drink and drive home? Have you always made the wise decision in related situations?

The Limits of Causal Reasoning

We can reason causally, and the fact that causal relations spring so easily to mind so often suggests that we frequently do. Examples like Simpson's paradox show that we have to reason causally. Now we will see that our causal reasoning abilities may be too powerful. They give us the ability to lie to ourselves in ways that may have short-term benefits but hurt us in the long term.

Have you ever failed to study adequately for a test? Laziness has the benefit of giving you time to do more pleasurable activities, but it has an additional benefit—it provides an excuse for failure. If you don't do well on the test, you can blame your failure on your lack of preparation. Not doing what is necessary to succeed is a form of self-sabotage and is not uncommon.[26] Procrastinating on work gives you the opportunity to tell yourself and others that you did not have sufficient time. Drinking until dawn or skipping a practice session lets you say that your poor performance on a test or in a sports match was not due to your lack of talent or ability but instead because you were tired or not at your best.

Self-handicapping has been demonstrated in the lab. In one study, students were told they would be taking an intelligence test and that a short practice session beforehand would facilitate performance.[27] Subjects were tested in one of two conditions. In one, they took the practice session, and in the other they were induced not to. The results showed that those who did not practice more often attributed poor performance on the intelligence test to the lack of practice. The results also showed that not practicing protected the students from feeling bad about themselves following a poor performance. It was only when students did practice that blaming themselves for their performance affected their self-esteem.

Self-handicapping is not recommended. Its consequences include studying less, getting low grades, drinking too much, and being unpopular because of the potential of becoming a "whiner."[28] It involves placing barriers to success in your own path, and that can be self-defeating. On top of that, those barriers lower your expectations for yourself both now and in the future.[29] One cannot, however, just assume that someone who blames external conditions for their fate is a whiner. After all, external conditions can be decisive. If you really did do your homework and left it on your desk but your dog ate it, then it really is your dog's fault and not yours.

The key question to determine blame is whether you merely observed the external conditions or intervened to create them. If your friends drag you to a bar on your birthday and keep you up late drinking against your will, then your poor performance the next day really could be due to staying up late. But if you decided to go drinking on your own, then you are the one to blame; you handicapped yourself. And if you did it intentionally in order to have a ready excuse for poor performance, then it involves not just self-handicapping but self-deception too. Self-deception involves intervening in your performance by taking an action to change your own performance while not admitting to yourself that you are doing so. The behavior still serves as an excuse for poor performance, but the fact that you undertook the behavior voluntarily suggests that you doubted yourself and felt you needed an excuse, and so you created that excuse.

Self-deception occurs when you contort yourself to see your image in the mirror from the best angle, or when you lose a few pounds before going to the doctor and then gain them back after the visit. In both cases, you are taking actions with the purpose of giving a false impression. In the first instance, you are the victim of the false impression. Most people see you from a less-than-perfect angle, so you are deceiving yourself about what you look like. This has obvious benefits as it may be good for your self-esteem to believe that you are more attractive than you are. Indeed, believing you are more attractive may make you more attractive. But in the second example, deceiving the doctor (and yourself) about your weight can only get in the way of the doctor's ability to care for you.

Self-deception has also been demonstrated in the lab. In one study, experimenters told a group of women that pain tolerance was indicative of the presence of a chemical that determines skin quality later in life.[30] There were two cover stories. In one, they were further told that high endurance to pain is indicative of good skin in the future (the high-endurance condition). In the other, they were told that low endurance is indicative of good skin (the low-endurance condition). Then the experimenters did something cruel. They asked the women to tolerate a painful (but harmless) stimulus for as long as they could. A device was used to apply painful pressure to their fingers—a pressure that they could relieve at any time by pressing a button. Women in the high-endurance condition tolerated more pain than women in the low-endurance condition. In short, they changed their pain tolerance to support a beneficial diagnosis of their future skin quality. In addition, the women were asked how much effort they had made to endure the pain. Those in the high-endurance condition reported less effort than those in the low-endurance condition even though they must have made a greater effort as they endured more pain. Not only did they last longer, but they told themselves it was easy, reinforcing the desired conclusion that they really had greater pain endurance.

What this study shows is that people will manipulate their behavior to convince themselves of something they want to believe. On the one hand, the women in this study had to believe that they were not controlling their pain tolerance, otherwise they would not have been able to conclude that their pain tolerance was low or high (they had to believe they were observing their pain tolerance, not intervening on it). On the other hand, they had to be in fact controlling (intervening on) it, otherwise there would have been no difference between the high- and low-endurance conditions. These two facts—a belief in no control while there actually was control—constitutes self-deception.

Self-deception is an example of how people manipulate their awareness of their causal beliefs to serve their self-interest. A smoker might hide the belief from themselves that they are unable to stop smoking. They might tell themselves that they have control and can quit anytime, even though their craving to smoke is sometimes overwhelming. A different form of self-deception arises in child abuse. In that case, the perpetrator might convince themselves that they have no control of the situation. The problem is not their own behavior; it is the child's behavior. In both instances, people are deceiving themselves about what is actually controlling their behavior in order to come to a desirable conclusion.

How to Reason Causally

It turns out that people with gray hair tend to earn more money than people whose hair is not gray. Therefore if your hair is not gray and you like money, you should dye it gray.[31] Hilarious, right? Maybe not. Still, it is obviously silly reasoning. The reason that gray hair is correlated with earning money is not because hair color is a cause of how much people earn; it is because both hair color and earnings are affected by growing older. While various biological mechanisms turn hair gray, other mechanisms increase salaries.

But it is not just the causal structure that makes the reasoning silly. It is also because dyeing hair is an intervention, not an observation. If you observe two people, one with gray hair and the other a different color hair, then a reasonable guess is that the person with gray hair earns more than the other one. Yet making your hair gray involves changing the color of your hair intentionally. As a result, whatever processes normally occur over time, they are no longer determining the color of your hair. You are. So dyeing your hair disconnects it from those normal processes, rendering it independent of your earnings. The surprise here is that your reasoning system knows all of this already. Perhaps you could not have articulated it, but you were reasoning as if it were true. Humans are sophisticated causal reasoners.

Despite human sophistication, we often fail to reason, not because we are lousy reasoners, but because we don't know enough relevant facts. Ask yourself how a ballpoint pen works, or a zipper, or a toilet. Can you explain how these objects work in any detail? Most people cannot. And these objects are relatively simple. What about more complex artifacts like iPhones or cars? Do you know how these work? A few of the engineers among you might have some sense, but even your knowledge is probably quite specialized. And the rest of you likely can only explain the mechanisms of operation of these things at an extremely superficial level. What about biological entities? Can you explain how a muscle works, how it uses energy to contract and relax, the chemicals that are involved, and its relation to the nervous system? What about a housefly? How does it manage to escape fly swatters so effectively? What about social entities like corporations, banks, and the judicial system? How much can you say about their inner workings and how they accomplish what they do? What about policies? Do you know how welfare works? What about the tax system? The world is so complex, and we know so little about how it works. There is just too much to know.[32]

Most of us are unaware that we know so little. This is the knowledge illusion—that people think they understand things better than they do.[33] My friend and colleague Phil Fernbach and I have argued at great length elsewhere that the reason we live in a knowledge illusion is because we fail to distinguish what we know from what others know. Others know how the judicial system works, so I don't have to. We each have our few areas of specialized knowledge and depend on others for the rest. That is how humans evolved in order to distribute cognitive labor across the family or tribe and get things done as a group, not individually. That is how our ancestors hunted, cooked, built boats, and did pretty much everything. Individuals had their specialties and by collaborating to accomplish tasks, the community became much more than the sum of its parts. And that is how we still function today. Most of us operate by collaborating in small groups. And when we are not collaborating, we are outsourcing our cognitive needs to others. We might hire someone to fix something that we can't fix, pay someone to fly us somewhere because we can't fly, or just make use of someone else's theory that we do not fully understand or let them amuse us with a story whose details are beyond us. We are constantly taking advantage of knowledge and skills that sit in other people but are available to us because we share a rich culture.

Deploying sacred values is one way we take advantage of the community around us. Making a decision using a simple action rule acquired from our community ("do not pollute!") is a way to outsource the problem; our community makes the choice for us.

Collective understanding is maintained in multiple ways. Anthropologist Joseph Henrich tells the story of an epidemic among the Polar Inuit tribe that killed off so many older members, the tribe lost access to much of its technology. It could no longer build traditional weapons, snow homes, and kayaks. The knowledge about how to build and use these tools was lost. The knowledge had resided in the heads of the older generation, leaving the tribe to its own devices when these members died.[34]

Cultural knowledge of this form can also be housed in formal languages like mathematics. Mathematics embodies generations of insights that the community puts together in the form of well-developed theories that cannot be attributed to any one person. Even music houses a lot of collective knowledge about scales, instrumentation, rhythm, and much more.

A composer acquires a lot of knowledge from their culture and uses that knowledge to compose a particular piece. But much of a composition depends on knowledge that is held by others. Composers do not have to know how all the instruments they write for work. Few composers of orchestral pieces could even play every instrument. Composing music does not require it; that is what musicians are for.

Much of the cultural information that shapes our beliefs and opinions comes in the form of narrative. Fables and folktales embody anecdotes, and relate beliefs and values (both outcome and sacred values) about how to treat other people (don't cook them in an oven even if your house is made of gingerbread). Science fiction is a fabulous source of scientific facts and theories. The importance of narrative in shaping public policy and attitudes has been noted by researchers in many disciplines. In his classic 1922 book *Public Opinion*, communications and journalism guru Walter Lippmann mapped out the role of propaganda in World War I.[35] In the 1940s, psychologists uncovered the power of propaganda films in World War II.[36] Propaganda during times of war—and at other times as well—tells a simple story of one's own side as a group of innocent victims attacked by an uncivilized nation for no good reason. The authors conclude in both cases that the propaganda was critical for motivating the heroic efforts of the populations during wartime.

The frequency of propaganda and modern fake news shows that cultural knowledge is not necessarily an accurate representation of reality. But even when it is not, it can be useful. Henrich has discussed multiple examples of actions that are culturally prescribed for a good reason, but the reason has been lost or might never have been apparent. For instance, religious rituals have the benefit of uniting a community, but they are justified as a means of appeasing the gods.[37] Or a culture might use spices to flavor meat on the belief that the spices enhance the flavor of the food, when in fact the original benefit of the spice was that it made the meat safe to eat due its antimicrobial effect.[38] Narratives are not necessarily grounded in truth. They are grounded in the benefits they supply to their purveyors.

The role of narrative has also been carefully studied by students of public policy. The narrative policy framework offers a theory of the central properties of narratives like settings, characters, plots, and morals.[39] For instance, public debate about climate change rarely focuses on hard-to-understand scientific evidence; instead, it tells stories that promote or denigrate characters (like former US vice president Al Gore, a staunch advocate of climate change mitigation, versus the fossil fuel industry) by sharing anecdotes that reveal the characters' intentions and battles. In the field of marketing, it has long been recognized that products are not sold by spelling out virtues but rather by using them as props in microdramas that associate the products with the happy, sexy, successful people we would all like to be.

In sum, people are persuaded to make decisions not by evidence but instead by stories that associate options with the identity that the person aspires to. Every community is bound by its own values and beliefs that form its identity, made public in the form of a narrative. Persuasion often consists of associating a target idea with the narrative that glues together a community's identity. The linkages are strengthened by embedding the idea within an appropriate set of causal relations.

Consider the simple, abstract narrative promoted by the fossil fuel industry that economic development consists of good people working hard together to use their natural resources to support growth. One of the most critical natural resources is fossil fuels as they are able to generate the energy that will allow a nation to attain a new level of prosperity and comfort. This narrative makes a pair of causal assumptions—that the use of natural resources causes growth and that fossil fuels cause prosperity and comfort. Detractors of this narrative focus on different causal claims: that the fossil fuel industry is a set of bad actors motivated by wealth and power who try to hide a path to even greater prosperity and comfort that will not destroy the planet—namely, the development of renewable energy.[40] The causal claims in this case are that the desire for wealth and power are the causes of the fossil fuel industry's behavior and that they are causing the potential of renewable energy to be underappreciated. Causal claims such as these are the elements that bind narratives together and give them their power to persuade.

The beauty of narratives is their "truthiness."[41] They are able to represent some aspects of a situation without capturing all of them by strategically ignoring some causal relations. The fossil fuel industry's narrative is not entirely wrong. Fossil fuels have served as the engines of economic growth—certainly in the West since the Industrial Revolution. Fossil fuels were in fact the critical resource used to develop cars, trains, and jet airplanes. And the manufacturing industries that launched the twentieth century depended heavily on fossil fuels. But that narrative is incomplete. It fails to identify the actual motivation of the fossil fuel barons. They were not in it simply to make the world a better place. They wanted to make money, and as much as possible. And they did. It is also possible that modern industry could have developed during the same period using battery technology fueled by sun and wind. We will never know because the fossil fuel industry managed to preempt the development of battery technology. The other (green) narrative is at least partially accurate too.

How Not to Reason Causally

The freedom that narratives offer to pick and choose causal relations makes them a powerful as well as flexible way to express any perspective, and to engage, entertain, and educate. It also makes them exceedingly dangerous because they cannot be refuted. This is why the world has become such a mecca for conspiracy theories. From QAnon and the deep state to flat-earthers, to the idea that COVID-19 treatments are the real reason for COVID deaths, wild claims are made constantly, inevitably supported by some compelling narrative about the evil intentions of a cabal of powerful people.[42] The problem is that narratives allow so much freedom in their construction that they can wrap themselves around any inconsistent data. For instance, photos of a round earth do nothing for a flat-earther. After all, if a cabal of powerful people could fool us into believing that humans walked on the moon, they could easily fake a photograph of a round earth. Evaluating conspiratorial claims is especially difficult because some of them are true. There really was a conspiracy to, say, murder Julius Caesar a couple thousand years ago.

Conspiracy theorizing is especially pronounced during periods of great uncertainty. One theorist has suggested that conspiracy theories "bring the disturbing vagaries of reality under . . . control"—a view that has some evidence to support it.[43] I am writing this book during such a period of great uncertainty. The world is experiencing rapid and unpredictable change, inhabited by strong, contradictory perspectives. In my ideal world, society would be busy attempting to reduce uncertainty by using all the scientific tools and techniques we have at our disposal to try to converge on a common causal model. Instead, we are living in separate, intellectually closed communities, cut off from one another, with each of our tribes held together by a common narrative constructed to support our sacred values. This does not bode well.

7 Values (the Nonsacred Kind)

All decisions require an assessment of value no matter how they are made. Those decisions made with sacred values will focus on the value of an action, like whether or not an act of killing is acceptable. One relevant sacred value would be that you should never kill a human being regardless of any benefit that might be obtained by doing so. Alternatively, as a consequentialist, I might value the outcome that no human being is killed rather than the action of not killing. I might value that outcome a lot. What is special about outcome values? First, there is nothing preventing me from engaging in material trade-offs with the outcome value. If, say, I can save five people from death by killing one, then that seems a good reason to kill one.[1] But I cannot do that if I take the sacred value seriously.

A second unique property of outcome values is that they are compatible with uncertainty in a way that action values are not. In chapter 4, I introduced EU (expected utility) theory. The *U* in EU theory is a way to represent outcome values. EU theory says that we should make decisions by considering both the probability of an outcome and its utility. If one outcome of an option is bad (has a small utility value), we might still consider the option if the probability of that outcome is low enough. So even if I would really hate killing someone, an action is acceptable if the probability of that outcome is close to zero. That is why it is acceptable to drive. Even though there is a chance that I will kill someone by getting behind the wheel, the probability is low enough that it is worth it. Driving is harder to justify for someone who lives by sacred values. If not killing is sacred to them, they really should not drive unless driving is a necessity required by other sacred values, like getting to church on time.

Another argument a consequentialist can make for driving is that not killing may be important, but other things are important too, like getting places. Even if they really value not killing people, not being stuck at home is also important, and its importance justifies driving given how low the probability of killing someone is. Again, sacred values do not support this argument.

Finally, an outcome value is justified in a different way than a sacred value; the two values participate in different sorts of logic. Sacred values get their justification from a system of sacred beliefs. For example, we might hold the value that thou shalt not kill because it says so in the Bible. It is one of the Ten Commandments. In that case, the argument for the sacred value is an argument from authority, but it is one presumably consistent with a set of other sacred values we hold. Or we might take a deontological perspective seriously and derive the sacred value from some other more basic premise or premises. In either instance, the justification for a sacred value is a matter of logical consistency.

Outcome values can also come from authority. My mommy told me that killing leads to bad consequences and therefore I prefer outcomes that do not involve killing. The kind of consistency that matters for outcome values is that they obey the axioms of EU theory. For instance, my outcome values should be transitive. If I hate outcomes involving death more than I hate outcomes of theft, and I hate theft outcomes more than I hate the outcome of deception, then I should hate death more than I hate the result of deception.

My point is not that people can justify all of their values. I might not be able to justify either kind of value. My values might just be grounded in a feeling or intuition. I might be disgusted by the thought of killing and hence have the sacred value not to kill. Or I might suffer panic at the thought of having killed someone and thus give high value to the outcome of not having killed. My point is that if I am going to try to justify my values, I will justify sacred and outcome values in different ways.

Sacred and outcome values can be hard to distinguish. Is the social justice mindset a sacred value or consequentialist mindset? I think the answer is that it depends on how one is reasoning about social justice. A person who asserts claims about which actions are acceptable and which are not is treating it as a sacred value. A person who is describing which states of the world they consider just and which are not just is being a consequentialist. It is merely a matter of framing. But even though the two values can appear similar, how one frames an issue matters. Frames have consequences!

Previous chapters have focused on the psychology of sacred values. The rest of this chapter will outline what psychologists know about outcome values.

Back to the Normative/Descriptive Distinction

If EU theory expressed all there is to say about outcome values, then we would be done. But it turns out that the notion of utility is just a starting point. This was first pointed out by economist Maurice Allais at an economics conference in Paris in 1952. Allais, who subsequently won a Nobel Prize, gave the following two choices to the assembled economists:

A. a sure gain of $1,000,000

B. ten percent chance to win $2,500,000, 89 percent chance to win $1,000,000, and 1 percent chance to win nothing

If you are like Allais or me, you would take option A. Why risk losing a cool million for a mere 10 percent chance of winning $2.5 million? He also, simultaneously, gave the group the following choice:

C. eleven percent chance to win $1,000,000, and 89 percent chance to win nothing

D. ten percent chance to win $2,500,000, and 90 percent chance to win nothing

On this one, both Allais and I prefer the second option. Why give up a chance to win an extra $1.5 million just because it means a piddly 1 percent additional chance of winning nothing?

There is a problem here. EU theory allows us to assign utilities to outcomes in any way that suits our fancy, but we do have to be consistent. According to EU theory, the two choices that Allais and I (and maybe you) made are not consistent. We prefer A to B. Therefore because we are choosing by expected utility, $EU(A) > EU(B)$, we must choose the option with a higher expected utility. If we spell out $EU(A)$ and $EU(B)$, we get

$$U(\$1\text{M}) > 0.1U(\$2.5\text{M}) + 0.89U(\$1\text{M}) + 0.01U(0)$$

All terms are just numbers, so we can do simple algebra. Let's subtract $0.89U(\$1\text{M})$ from both sides of the inequality:

$$0.11U(\$1\text{M}) > 0.1U(\$2.5\text{M}) + 0.01U(0)$$

Now if we add $0.89U(0)$ to both sides, we get

$$0.11U(\$1M) + 0.89U(0) > 0.1U(\$2.5M) + 0.9U(0)$$

The left-hand side is just the expected utility of option C, and the right-hand side is the expected utility of option D. So we conclude that $EU(C) > EU(D)$. In other words, if we choose according to EU theory, then if A is preferred to B, C should be preferred to D. But it is not. Similar logic implies that if B is preferred to A, then D should be preferred to C. Anyone who chooses A and D, or B and C, is violating EU theory. Note that the proof allows any utility function at all. I did not even specify that $U(0) = 0$. The only requirement is that both choices involve the same utility function. In that case, EU theory places real constraints on what we should choose.

Allais viewed his response to the thought experiment as a mistake. He thought we should revise our choices to be consistent with EU theory. After all, if EU theory is a normative theory that dictates the best way to make a decision, then we should do our best to make sure our decisions conform to it.

But I am going to focus on a different, descriptive question about why people act the way they do. What explains the violation of EU theory exemplified by the Allais paradox in the first place? How are people choosing? The likely answer to that question is called the certainty effect. Option A gets a bump in value because it involves a guaranteed outcome. As we saw in the last chapter, uncertainty can be distasteful.

Framing Effects and Risk Attitudes

Here is a simpler example showing that people prefer to avoid uncertainty:

> Assume yourself richer by $300 than you are today. Choose between
>
> a. a sure gain of $100
>
> b. 50% chance to gain $200 and 50% chance to gain nothing[2]

Most people prefer the first option. They prefer a sure thing over a risky option with identical expected value. That is the certainty effect. Consider the following choice, however:

> Assume yourself richer by $500 than you are today. Choose between
>
> c. a sure loss of $100
>
> d. 50% chance to lose $200 and 50% chance to lose nothing

Now most people choose option d, the risky option. They prefer uncertainty when they are choosing between losses. As you might have noticed, this pattern of choices is not rational because option a has an identical outcome to option c. Option a has me giving you $300 and then giving you another $100, whereas option c has me giving you $500 and then taking away $100. In both cases, you are left with $400. Similarly, options b and d are identical in terms of outcomes (they both involve a 50 percent chance of walking away with $300 and a 50 percent chance of walking away with $500). So if you choose based on outcomes, you should either take a and c, or b and d. The fact that most people take a and d suggests that we prefer certainty in the domain of gains (as in the Allais paradox), but we prefer to be risky when we are choosing between losses.

People are risk averse in the domain of gains and risk seeking in the domain of losses. Uncertainty is distasteful when we are thinking about what we might get, but it becomes attractive when we are thinking about what we might lose. This is a framing effect. People do not make decisions based on actual outcomes as expected by EU theory. Instead, people make choices relative to a reference point that determines whether they are gaining or losing. Thus EU theory fails as a descriptive model—as a model of how humans behave.

Framing effects have been demonstrated in a variety of situations. Here is a demonstration in the context of negotiation:

A large car manufacturer has recently been hit with a number of economic difficulties and it appears as if three plants need to be closed and 6,000 employees laid off. The vice president of production has been exploring alternative ways to avoid this crisis. She has developed two plans:

Plan A: This plan will save one of the three plants and 2,000 jobs.

Plan B: This plan has a 1/3 probability of saving all three plants and all 6,000 jobs, but has a 2/3 probability of saving no plants and no jobs.[3]

These two options are framed in terms of gains; they concern how to save jobs. Accordingly, over 80 percent of people asked chose plan A, the risk averse choice. Another group was shown the following two options, framed in terms of losses:

Plan C: This plan will result in the loss of two of the three plants and 4,000 jobs.

Plan D: This plan has a 2/3 probability of resulting in the loss of all three plants and all 6,000 jobs, but has a 1/3 probability of losing no plants and no jobs.

Over 80 percent of this group chose plan D rather than plan C. This is remarkable given that plans A and C are identical, and so are plans B and D. The two pairs of options represent identical choices. Framing that choice differently reversed what people chose.

It is important to be careful when interpreting results like these. For instance, participants might understand the statement of the problem in unexpected ways. In this case, saying that 2,000 jobs will be saved or 4,000 jobs will be lost is slightly ambiguous. It could be read as implying *at least* 2,000 jobs will be saved or *at least* 4,000 will be lost.[4] If so, that could be why people choose options A and D over B and C, respectively. I don't think that this sort of alternative understanding of the question explains the effect. My confidence comes from two places. First, even though I understand the statements as intended, I still feel the pull of the certain option in the gain frame and risky option in the loss frame. So my intuition supports the reality and irrationality of the framing effects. Perhaps yours does too.

A more convincing argument comes from a study published in Science in 2006.[5] In that study, subjects were asked to imagine that they received $50 and then made a choice. In one condition, that choice was between keeping $20 of the $50 or spinning a wheel that was ⅓ red and ⅔ green. If they chose the gamble—the wheel—then they would be able to keep the entire $50 if the spinner landed on red or otherwise they would lose it all. A second condition was exactly the same except that the first option was not to keep $20 but instead to lose $30. Keeping $20 of the $50 is exactly the same as losing $30 of the $50, so the outcomes in the two conditions were indisputably identical. Yet there was a large framing effect. I conclude that it is true that people tend to be risk averse in the domain of gains but risk seeking in the domain of losses, and this effect arises even when outcomes are identical, just described differently. This poses a serious challenge to the idea that people choose according to EU theory.

Framing effects need to be taken seriously. Valerie Reyna, a prolific judgment and decision-making researcher at Cornell University, has shown not only that life experience does not diminish framing effects but that they may even get stronger over time as well. Moreover, they affect judgments by experienced experts even when people's lives could be at stake. Reyna and her colleagues tested Cornell undergraduates, college graduates, and a group of intelligence professionals from a federal agency by having them make choices between certain and risky options under gain and loss frames.[6] The federal intelligence agents showed more framing effects than both the undergraduate students and postcollege adults, even on matters of life and death.

Loss Aversion

Gain versus loss frames do more than determine people's risk preferences. They also affect how we respond to potential outcomes. In particular, we care more about potential losses than potential gains. Would you accept an even bet of $100 on a flip of a coin (heads I owe you $100, and tails you owe me $100)? Most people would not. This might be due to an aversion to risk. But another explanation is that it is because the prospect of losing $100 hurts more than the prospect of winning $100 gives pleasure. People are averse to losses more than they are attracted by gains. In a famous line from Danny Kahneman and Amos Tversky, "Losses loom larger than gains."[7] Loss aversion explains why people pay for warranties and insurance policies that just aren't worth it (and why companies make so much money selling them). Why pay $39.95 for an extended warranty for a TV that will be outmoded in a few years anyway? Why pay for rental car insurance every time you rent a car when you so rarely take advantage of it? After all, if these were good deals for you, then the companies wouldn't offer them. The answer is that people will pay to avoid the prospect of loss. Loss might still occur (that is what all the fine print in the contracts is about), but the feeling of being protected from downsides is worth a lot to people.

Imagine that you have lung cancer and are choosing between surgery and radiation therapy. Here are some facts to consider:

Surgery: Of 100 people having surgery, 90 live through the post-operative period, 68 are alive at the end of the first year, and 34 are alive at the end of five years.

Radiation Therapy: Of 100 people having radiation therapy, all live through the treatment, 77 are alive at the end of one year, and 22 are alive at the end of five years.

When researchers gave groups of actual patients, business students, and experienced physicians these options, 82 percent chose surgery. The researchers also gave a different group of subjects the same facts under a different frame:

Surgery: Of 100 people having surgery, 10 die during surgery or the post-operative period, 32 die by the end of the first year, and 66 die by the end of five years.

Radiation Therapy: Of 100 people having radiation therapy, none die during treatment, 23 die by the end of one year, and 78 die by the end of five years.[8]

Under this frame, only 56 percent of the subjects chose surgery. At this point, I am sure you have figured out for yourself that the options are identical in the two cases. The only difference is that the first two options present the facts in terms of how many people live so we can call it a survival frame. The second two describe the facts in terms of how many die, so it is called a mortality frame.

Why is surgery more attractive under a survival than mortality frame? I suspect the key difference is the survival frame's innocuous "90 live through the post-operative period" versus the mortality frame's threatening "10 die during surgery or the post-operative period." Who wants to choose a procedure that might kill you in the near future? This is loss aversion at work. Losses loom large—much larger than the potential benefits.

What should medical professionals do when they describe treatment options? Should they use a survival frame or excite loss detectors with a mortality frame? Loss aversion can be a tool to suggest the procedure that one thinks best, but what if you want to leave the choice entirely up to the patient without suggesting an answer? One answer is to provide both frames to ensure that you are not introducing any bias. This is not a great solution, though. Its main effect is to introduce confusion by inducing the patient to think about the problem in ways that conflict with one another. It would be like overhearing the statement "Pat is running" and then having someone clarify by saying, "Pat is a politician up for reelection who likes to run marathons." No help at all.

The prospect of loss hurts. That is why people remain in living situations, relationships, and jobs that make them unhappy. Sure, imagining all the benefits of changing your situation is appealing, but imagining the prospect of losing what you have overwhelms those benefits. Loss aversion also serves an important social function. It maintains the status quo. If losses loom larger than gains, then people will tend to be conservative because they would rather not suffer the costs of change even if it means missing out on the benefits. Referenda that propose to continue or strengthen existing policy are voted in favor of by citizens of a community more often than referenda that change existing policy.

Loss aversion prevents people from trading too. Researchers have found that people will demand more money to sell something they own than they are willing to pay for that same thing if they do not own it—a finding called the endowment effect. The original experiments showing the endowment effect set up a trading situation. People were either endowed with a mug or a few dollars. They found little trading because, on average, people were not willing to pay as much for the mugs as was demanded by the mugs' owners.[9] In a different study, shoppers were either taxed 5¢ for using a disposable bag or offered a 5¢ bonus for using a reusable bag. The tax made shoppers much less likely to use disposable bags whereas the bonus had little effect.[10]

The endowment effect may result from the fear of getting a bad deal, buyers and sellers being strategic in their offers, or the mere fact of ownership.[11] The influence of mere ownership was shown in a clever experiment. People who already owned a mug were offered a second identical one. They were willing to pay more for it than those who did not own such a mug.[12] So ownership makes things more valuable. In fact, merely imagining owning the good also makes it more valuable.

If what I have has more value just because it is mine, and the same goes for you, then bargaining will be difficult. I will not be willing to, say, give you land if you reduce the number of missiles in your arsenal because my land is worth more to me than it is to you and your missiles are worth more to you than they are to me. One might hope that these effects would dissipate over time and get smaller if you imagine trading not today but instead in the future, such as a year from now. One's hopes would be dashed. The effect just gets stronger when imagining the future.[13]

What is the source of loss aversion and the endowment effect? Do they result from the influence of framing on how people feel or what they think about? There is some direct evidence that they reflect a hedonic response—that people show a stronger affective response in the face of a loss than in the face of a gain. The evidence comes in the form of a stronger signal in areas of the brain involved in affect when buying rather than selling.[14] Another source of the effect appears to be that people attend to dimensions of decision options that are consistent with the frame they are choosing from. For instance, buyers focus more on negative attributes of products, whereas sellers are more focused on positive attributes.[15] In sum, both feelings and thoughts seem to be culpable.

Whether the effect describes how people emotionally react to gains versus losses as opposed to what they think about under the different frames, the effect presumably has a deep evolutionary source.[16] The world offers costs and benefits, but the potential costs are more extreme than the potential benefits. The ultimate cost is death, and there is no corresponding ultimate benefit. Perhaps the ultimate benefit would be stumbling into the garden of Eden and having all of your cravings fulfilled. That would be pretty good, although it does raise questions about the dangers of getting all that you wish for. After all, why go on living if all of your aspirations are satisfied? Beyond that problem, however, the garden of Eden is still not the mirror image of death because you might have a heart attack the day after arriving and not be able to enjoy it. There is just nothing we can gain in life that is of the same magnitude as the things we can lose: our own life, and the lives of our family members and other members of our tribe. There is a deep asymmetry between gains and losses in the environment, whether we are talking about today's environment or the one that humanity evolved in. So it seems reasonable that we would evolve to fear losses more than we are enticed by gains. We have more to lose than we have to gain.

Loss aversion is important in all sorts of judgments. For one, it influences our sense of fairness. Consider this example:

> A company is making a small profit. It is located in a community experiencing a recession with substantial unemployment and inflation of 12%. There are many workers anxious to work at the company. The company decides to increase wages and salaries only 5% this year.[17]

Is this unfair? Most people say no. It is tough economic times, and the company gave everyone a reasonable raise of 5 percent. Now consider this case:

> A company is making a small profit. It is located in a community experiencing a recession with substantial unemployment but no inflation. There are many

workers anxious to work at the company. The company decides to decrease wages and salaries 7% this year.

Is this unfair? Definitely. Who can tolerate a decrease in their earnings? Note, though, that a 7 percent decrease when there is no inflation reduces buying power to the same extent as a 5 percent raise when there is 12 percent inflation. But taking away is perceived as unfair regardless of the real consequences because losses loom so large in our judgment.

Prospect Theory

In chapter 4, I discussed a simple, elegant normative theory of consequentialist decision making, EU theory. Is there a corresponding simple, elegant descriptive theory? Can we describe how people make decisions in a way that everybody can understand? No, we cannot. People have to make too many different types of decisions, from decisions about whether or not to inhale, to whether or not to visit grandma, to whether or not we should buy a home whose location might make it vulnerable to the vicissitudes of climate change. Each of these decisions must be made with different amounts of time pressure, depend on different kinds of knowledge, require different degrees of consultation with experts, and elicit different kinds of principles and values. No one theory will ever be able to accommodate all of that.

The best we have is another insightful contribution of Tversky and Kahneman called prospect theory. Though the theory does not explain everything, it represents several of the key ways that people deviate from EU theory in an exceptionally clear and helpful way. What I find most useful about it is the clear way it distinguishes loss aversion from framing effects.

Prospect theory posits that people approach decisions in two phases. First, they frame and edit the decision. This includes determining a reference point. For instance, are they already endowed with one of the options or not? Reference points are constructed by the way the decision is presented, along with the decision maker's expectations influenced by their understanding of norms. In New York City, for example, they should expect the price of a cup of coffee to be higher than in Providence, Rhode Island. Reference points are also influenced by habits. If it is my habit to have a cup of coffee in the morning, then my reference point will include having that coffee.

In the second phase of prospect theory, the decision maker is assumed to evaluate individual options. Prospect theory focuses on simple gambles consisting of a probability of winning a particular amount of money. It (like EU theory) assumes that for each gamble, people represent its uncertainty and its value, and multiply them to determine the value of the gamble (called a prospect). They do this for each gamble being offered and, again like EU theory, choose the highest. The theories differ in how uncertainties and values are determined.

Let's start with value. Prospect theory assumes a value function that assigns a value (v) for each possible monetary outcome. This value function has three important properties. First, it is defined on gains and losses. Utility functions are not; they are defined in terms of total wealth starting from 0. But the value function has a point in the middle where the value assigned is 0, meaning $v(0) = 0$. This is the reference point, and values are determined in terms of how much they deviate from this point. All monetary outcomes on the x-axis to the right of this reference point have positive value, and everything to the left is a loss and has negative value. The second property of the value function is that it is concave (leaning forward) for gains, but convex (the opposite shape) for losses. We saw in chapter 4 that concavity implies risk aversion for utility functions and exactly the same logic holds for the value function. So the curve represents risk aversion in the domain of gains. Convexity, in contrast, implies risk seeking, so the function implies that people should be risk seeking in the domain of losses. This is exactly the pattern we saw with framing effects. Finally, notice that the curve is steeper for losses than for gains. This is how the theory represents loss aversion. Because the loss curve is steeper, the value of a loss of any size is more negative than the value of a gain of the same size is positive; the absolute value of $-x$ is greater than the absolute value of $+x$ for any outcome x. Losses loom larger than gains.

On this curve, framing effects—risk aversion in the domain of gains and risk seeking in the domain of losses—are represented by the fact that the gain function bends down (is concave) but the loss function has the opposite shape (it is convex). In contrast, loss aversion (the fact that losses loom larger than gains) is represented by the difference in slopes between the two curves. The two concepts have entirely different representations in prospect theory.

The other key element of prospect theory is called the π function (π is pronounced "pie"). It takes a probability as input and gives a decision weight as output. A decision weight refers to how much influence the corresponding outcome has in determining the overall value of the prospect. For instance, if the probability is 0, then the π function assigns a decision weight of 0, $\pi(0)=0$, because outcomes that have 0 probability should be ignored, and are ignored according to prospect theory. On the flip side, $\pi(1)=1$. Probabilities of 1 are given all the weight in a decision—a weight of 1. Other values vary some. For low probabilities, $\pi(p) > p$. That is, for probabilities obviously above 0 yet below about 0.2, the decision weight is greater than the actual probability. This is called overweighting. The idea is that we pay too much attention to low-probability events like an important phone call coming in when we take a shower, the puppy destroying the house if we go out for dinner, or winning a prize at the local talent show. These are events that do happen, and we should be influenced by them, but we tend to allow them to influence us too much.

According to prospect theory, we treat very low probabilities differently. The decision weights for these are unstable and that is why the weighting function is absent (undefined) for probabilities extremely close to 0. Sometimes we overweight them. For instance, terrorist events are tragic when they happen but they almost never do, at least in the United States. Yet many of us are concerned about them, especially when we get on an airplane. Winning the lottery is another example. The probability of winning the jackpot in a state lottery is akin to that of getting hit by lightning more than a hundred times over the course of a lifetime. Nevertheless, some people play the lottery on a regular basis. But sometimes we neglect very low probabilities altogether. We don't bother ensuring that the lighting fixture above our beds is on tight so that it doesn't come crashing down while we are sleeping. And when we go for a stroll in the local park, we don't carry a rifle in case a big cat has escaped from the local zoo. We don't worry about most very low-probability events—a sign that we can be quite rational.

Very high probabilities are also unstable. We are drawn to certainty and tend to treat very likely events as if they are certain. We assume the sun will rise tomorrow morning. If you're anything like me, you assume that you will want breakfast tomorrow morning. But some events that are close to being guaranteed are not treated that way. Some of us worry that the airplane we are flying in will not get back to earth safely or fret that we won't arrive on time even if we give the trip twice as much time as it normally requires.

Events with probabilities that are in the middle of the range—not too high or too low—tend to be given less weight than they should. This may be because we fail to consider all the ways they might occur. If I judge the likelihood that poverty in Asia will be relieved, it might escape my notice that the probability is higher than I thought because the size of the population is declining. If I am asked about the probability of precipitation, I will think about the possibility of rain or snow while neglecting the possibility of drizzle or sleet.[18] This is one instance of a broader property of human cognition. Because both our memories and imaginations have limited capacity, we tend to be myopic. We fail to consider things that we should more than we consider things that we should not.

Preference Reversals

The picture that certainty effects, framing, and loss aversion paint is that people differ in systematic ways from what EU theory dictates. The very notion of utility has been questioned by some. The great University of Michigan philosopher Elizabeth Anderson argued in 1993 that human preferences cannot be reduced to a single notion of utility.[19] Rather, values are pluralistic. We respond in a large variety of ways to events, people, and outcomes. We might think of the same person as attractive, informative, delightful, honorable, corrupt, dangerous, and so on. Each response has its own time and place and cannot be packaged as a single number called utility.

The same point has been made in a less dramatic way by psychologists Paul Slovic and Sarah Lichtenstein. They offered participants a choice between two bets like the following:

H: 8/9 chance to win $400.

L: 1/9 chance to win $4,000.

Most people chose the H (high-probability) bet. Next, they asked each subject to price each gamble by stating the lowest price at which they would be willing to sell each gamble if they owned it.[20] What they found was that the strong majority of their subjects priced L above H. Even though they preferred H when choosing, they demanded more money for L. On the assumption that how much money one demands for an option is another measure of preference, this is called a preference reversal. Choice gives one order of preferences, but pricing gives the opposite.

Slovic and Lichtenstein went on to show this effect multiple times, including in a casino for real money. They pointed out, like Anderson, that the results challenge the whole concept of utility. Utilities are just preferences in disguise. They reflect how willing we are to decide in favor of an option. But it looks like our willingness depends on how the question is asked. Ask someone to choose, they say one thing. Ask them to rate its value on a scale like money amount, and they say the opposite. This is not so different from Anderson's point that values are pluralistic.

Why do preferences reverse? Choosing and pricing elicit different strategies. When we make a choice, we like to be able to justify it to ourselves and others. So we focus on a single dimension, the one that is easiest to justify. In this case, we pick the option with the highest probability—the H bet—because it is simple and compelling to pick the option that offers near certainty of a good outcome.

When pricing, though, this strategy is not available because the price must reflect not just probability but also how much you might win. So people use an anchoring and adjustment heuristic. This heuristic involves anchoring on a numerical value and then adjusting to improve the estimate. What should the anchor be? The subject wants to end up with a monetary amount, a price, so it makes the most sense to anchor on the monetary amount of the bet. For the H bet, this amount is $400, but for the L bet it is $4,000. The anchor is much higher for the L bet. It turns out that adjustments are typically insufficient.[21] People do not adjust as much as they should. In this case, they have to adjust to a lower value in order to account for the probability of losing and getting nothing. Because they don't adjust enough, they end up with a price that reflects the monetary amount more than the probability. So the L bet gets a higher price.

This hypothesis is a special case of the compatibility principle in the study of how people interact with machines. Do you prefer vehicles with steering wheels that you turn left to go left or turn right to go left? Answers to this question are pretty much universal because we like our actions to be compatible with the response we desire. Similarly, when operating with a decision option that has multiple dimensions (like probability and monetary amount), people give more weight to the dimension that is compatible with the response that they are asked for (like monetary amount when asked for a price).[22]

The upshot is that decisions are based on various subtle psychological processes; people are not simply maximizing utility. EU theory assumes that utilities are known before a choice is made. Economists often embody this assumption in the dictum that utilities are elicited. Yet the kind of work that I have discussed implies that values are not elicited; rather, they are constructed when the preference question is asked.

This conclusion irked a couple of economists, David Grether and Charles Plott. They appreciated the challenge this work posed to economics as the notion of utility is fundamental to much of economic theory. And they were skeptical of the work, so their plan was "to discredit the psychologists' work as applied to economics."[23] They offered thirteen methodological objections to demonstrations of preference reversals. These included the claims that participants in the studies were poorly motivated because so little was at stake, that some of the effects arose because subjects possessed different amounts of income in the different conditions because of the order of the two decision tasks, that subjects were actually indifferent to the options and were posturing, that subjects were mostly psychology students and hence confused, and that the experimenters were psychologists, so nobody trusted them.

To remedy these problems, they ran their own study. In their minds, they fixed all the problems of the psychologists' studies. They tested only economics and political science students. They used real money and careful instructions. They randomly varied the order of tasks and gave subjects a way to say, "I don't care which option I get." They used a special procedure to elicit prices. Finally, they themselves (trustworthy economists) conducted the experiment.

The result of this attempt was larger preference reversals than the psychologists had found. Given all the effort they made to show the effect was not real, I take this as strong evidence that preference reversals are in fact real, reliable, and robust.

Mental Accounting

We have seen that we can reverse people's preferences by framing options differently (in terms of gains versus losses) and posing decision problems differently (in terms of choice or pricing). It is hard to see how these manipulations should affect the utilities adopted by EU theory as utilities are generally assumed to be fixed to outcomes, and outcomes are identical in the various conditions of the studies described. As long as the outcome does not change, the utility should not change. One could argue that the manipulations do change the utilities, but then utilities are not just outcome values; they depend on how options are described and questions are posed too. It is no longer clear what a utility is other than a description of what someone chooses. The theory becomes circular. In sum, both framing effects and preference reversals pose a serious challenge to utility theory.

The concept of utility has been challenged in another way as well. Imagine George and Georgette, whose jobs are identical as is their performance. Moreover, they have the same savings and expenses, and so are in exactly the same financial situation except that George, being a man, earns $1,000 more a year than Georgette. They both would like to pay down their student debt but they both also have their eye on a fabulous pair of electric boots that are outside their normal price change. This year, Georgette was surprised when her boss gave her a $1,000 bonus. So this year, George and Georgette have exactly the same income. Who is more likely to pay down their debt and who is more likely to buy the boots?

This is an issue of mental accounting.[24] Money is so useful because it is fungible; it can serve as a substitute for so very much, more than anything else can. People will trade most things for money. Yet money is not always treated that way by consumers. George's extra $1,000 goes into his overall wealth budget because it is part of his normal salary. But for Georgette, the money is a windfall and goes into a special mental account. Windfall accounts can pay for surprise luxuries like electric boots. After all, the money is unexpected so it can be used frivolously. In contrast, wealth has to be treated with conservative respect, and so George is more likely to pay down his debt. Relatedly, when people received a coupon for $10 off at a grocery store, they spent on average $1.59 more on groceries.[25] The store not only lured them in but also sold them more. And it sold them more unique items. Customers who received the coupon were 5 percent more likely to buy items they had not bought before and would not buy again. Windfalls open people's pocketbooks.

Mental accounts are useful for helping people to organize and manage their budgets. The obvious application is for people on a limited budget to control their spending (20 percent on rent, 30 percent on food, etc.) or for anyone to control their spending when they go to a casino. A good idea when gambling is to arrive with only a limited amount of cash and no other way to pay. That will prevent you from gambling away your savings.

Mental accounts are not only for money. They can, for example, apply to food. You might budget where your calories are coming from (make sure the percentage coming from carbs is not too large). Or you might budget your time, only allowing yourself, say, an hour of mindless videos per day.

The sunk cost effect is a product of mental accounting. A good way to get someone to invest in your project is to first get them to invest a little, because once someone has sunk money into a project, they are more likely to sink more even if they are chasing good money after bad. Robert Moses was an urban planner who is as responsible as anyone for the development of the highway infrastructure in and around New York City. One of his principal strategies to get the NY legislature to fund his projects was to get it to invest just a little as a pilot project. Once he had his foot in the funding door, it was easy to convince legislators to spend more and more until the project was complete.[26]

Sunk cost effects arise because once one has tapped into a mental budget, there is incentive to pay out the rest of the budget. Stanford Graduate School of Business professor Chip Heath showed this in 1995.[27] He asked people whether or not they should invest further in hypothetical real estate projects that had fixed budgets that they had already invested in. If they did not invest more, they could use the money to invest in other projects. What he found was a sunk cost effect. People pursued projects they had already invested in even though they would have earned more by giving up and investing elsewhere. But the real kicker is what happened when subjects exceeded their investment budget. Now Heath saw the opposite of a sunk cost effect. People could have earned more by investing more but they didn't because they had emptied their mental account.

Mental accounts are certainly useful in helping us control our spending, limiting it when we are in danger of spending too much, and helping us see when we should be spending more. In that sense, they are entirely rational. But they are not rational according to the high standards of utility theory. Money is fungible, and an optimal decision maker would always spend the right amount without taking psychological factors into account. Surely it is more rational in a broad sense to take human limitations into account than to try to always give an optimal response.[28]

Constructive Preference

This chapter has reviewed some classic work on the psychology of decision making. One of the main lessons of this work is that people do not always know what they want. Much of the process of decision making is figuring out what your own preferences are. This makes people much easier to manipulate. A good publicist or advertiser will tell us what we want and make us believe it. Most of us did not realize we wanted to walk around with a big brand name emblazoned on our clothing until a clever marketer hired by the company sold us on the idea.

The fact that preferences are constructed places a real limit on the power of consequential decision making. If outcome values are constructed at the time of judgment, then good decision making should not necessarily optimize those values. They are too arbitrary and under the control of outside interests. But when we do know what we want and thus are not vulnerable to framing effects, preference reversals, and mental accounts, then it seems to make sense to strive for the best consequences that we can. After all, that is by definition how to get the most of what you want out of a decision.

8 Simplifying Choice

Imagine you had an app on your smartphone that could make all of your decisions for you. You told it what needed to be decided, and it was guaranteed to make at least as good a decision as you would have made yourself. No longer would you have to fret about what to wear, whether or not to send that difficult email, or who to vote for. You could outsource the choice to your phone, and it would make as good a call as you could. Such an app is science fiction but it is not dreamy, magical science fiction that breaks all the rules. I'm not suggesting the app could see into the future and always deliver the best outcomes. My science fiction app only knows what you know or could know by researching the question on the internet. So its knowledge is limited in the way yours is. Given that, it is a solid adviser.

An app like this would be the ultimate sort of simplifier. Instead of deliberating, you could just outsource your decisions. This would give you much more time for pleasant tasks like baking brownies, solving a Millennium Prize math problem, or whatever suits your fancy. A few of you might never use the app because you actually enjoy decision making. You are the exceptions. Most people loathe decision making, especially when it is hard and the consequences matter.

Whether or not to use such an app would be a complicated issue. After all, if an app is making your decisions for you, then you don't get credit for the decision. But you don't get blame either. This can be problematic. What if all the US Supreme Court justices used the app to decide on the future of capital punishment? Even if you accepted that the app was wiser and more informed than a Supreme Court justice, you might feel that an app should not be making important decisions for the country. The justices themselves should take responsibility for those decisions. Similarly, if you decide to end a close relationship, you might want to feel that it was your decision and not an app's. Rest assured that the person you are ditching would agree.

Nevertheless, there is an allure to outsourcing decisions. That is why we do it all the time. Pilots and taxi drivers choose our routes, chefs and product managers choose our ingredients, and legislators choose our laws. In the influential book *The Paradox of Choice*, Barry Schwartz argues that we would all be better off if we made fewer decisions or at least did not work so hard to make the decisions that we do make.[1] Schwartz maintains that there are two types of people, maximizers and satisficers. Maximizers strive to make the best possible decisions. They spend time and effort investigating options, determining outcomes, and thinking about the trade-offs involved. In contrast, satisficers choose the first option that meets their minimal criteria. They spend less time and put out less effort when making decisions. Maximizers generally end up with better outcomes, but it is satisficers who end up happier.

Schwartz's starting point is that those of us who live in the West are deluged with options and choices. Not only can we choose from hundreds of brands of breakfast cereal and thousands of phone plans, but nowadays we can choose our family arrangements and even how we present to others. We can modify who we hang out with, what we wear, and—to some extent—even our physical appearance. Never have we had it so good. Our ancestors could not even dream of the smorgasbord of options we have available every day. It was not long ago that most people were born into a family that pretty much determined where they lived, what they believed, what they were able to eat, and who they married. Most of us regard such a lack of freedom and control with horror. Yet it has its virtues. It frees us from decision making. Putting time and effort into making decisions is correlated with less satisfaction as well as more frustration with the outcomes we get. Making decisions requires an unpleasant kind of cognitive effort, the effort of figuring out both what you want and what each option is likely to deliver.

If you want to be a maximizer in this environment, careful decision making requires that you research the options available. This introduces you to all the options that you will not receive. You will learn all about the phone plans you did not choose. Even though the options you do not choose may be worse overall than the one you do choose, they are pretty much guaranteed to be better in at least some respect. So by discovering and thinking about those options, you are learning about what you are not getting, and this leads to regret.

This "tyranny of choice" caused by maximization may be largely a Western phenomenon. It turns out that in China, maximizers and satisficers have similar senses of well-being.[2]

The upshot is that, at least for Westerners, minimizing the amount of time we spend making decisions—at least those whose consequences are not critical—can be healthy. We can do that by simplifying choice. Satisficers do that by trying to meet only moderate expectations and accepting less-than-perfect outcomes. In fact, all of us use a range of strategies to simplify choice.

The Simplicity of Action Rules

A common way to simplify choice is to dispense with consequentialism by avoiding all the difficult reasoning and soul-searching that it requires. Instead, focus on actions and set up rules to act by. A simple such rule is *choose by habit*. As Barack Obama told the magazine *Vanity Fair*, "You need to focus your decision-making energy. You need to routinize yourself."[3] He makes it a habit to get to the gym at 7:30 every morning. He wears only blue or gray suits. He refuses to devote mental energy to making these decisions on a regular basis. He simply does the same thing whenever faced with a choice. Do not feel bad if you order the same item every time you visit a particular restaurant, go to the same place every year on vacation, or always sit in the same chair in the living room. You are just simplifying your decision making, and you are in good company.

Less simple decisions can also be governed by action rules. That is why we so often defer to sacred values. Sacred values allow us to choose without thinking. It is much easier to always avoid eating meat than to calculate at every meal the amount of animals' pain and suffering associated with each meal option. And it is much easier to simply never drive a gas-guzzler than to calculate the many costs and benefits to the environment of each transportation option every time you have to get somewhere. So behaving according to such sacred values is sensible. Even better, living by sacred values is frequently condoned by the community around us.

Sacred values have other benefits too. If there is a dispute in your community, bystanders generally follow the rule "choose the side of the disputants whose actions are most morally acceptable."[4] This rule works because actions are a public signal about intentions and the willingness to exploit others. Other signals, like outcomes and disputants' identities, are not always publicly available and therefore are less useful. Bystanders use their community's sacred values to identify which actions are acceptable.

Sometimes action rules cause us to oversimplify, becoming more of a nuisance than a help. One form of oversimplification is to get stuck in our actions. Telling the same story over and over, for example, means you don't have to come up with new, untested stories, but it can be less than stimulating for your friends and family. Always taking the same route to work regardless of traffic conditions and construction work obviates the need to decide which way to go, but it can make you late and frustrated on occasion.

Behavior changes when we come to think that there is only one result that matters, the one implied by our action rule. An extreme version of this linkage is called means-end fusion. According to distinguished University of Maryland psychologist Arie Kruglanski and his colleagues, actions can be perceived as an end in themselves. The action (a means) becomes fused with its goal (or end).[5] For instance, sometimes we eat just for the sake of eating or stroll for the sake of strolling. Fusion is also more likely to occur when the action and goal have been paired in the past and are similar. For example, people are more likely to practice a skill if their goal is to perform well than if their goal is to play a game because practicing and performing well are more similar than practicing and playing.[6] Practice eventually becomes fused with performing well and becomes habitual.

Means-end fusion changes the nature of an action. The action becomes especially rewarding because it is equivalent to goal attainment. If work and success are fused, then working itself is rewarding because it has all the rewards of success. And if smoking and satisfaction are fused, then smoking itself becomes rewarding. The action comes to elicit the same emotional responses as goal attainment.

Because actions that are fused with their goals are more rewarding, they inspire greater commitment. I am more likely to work and work longer if working gives me all the benefits of success. Sacred values also inspire greater commitment, perhaps for the same reason. In fact, sacred values could represent a type of means-ends fusion. If my sacred value is to minimize my carbon footprint, then any action I take that reduces the amount of carbon I generate is done for its own sake. Or if I believe in a constitutional right to carry arms, then carrying a firearm is its own purpose. The action is interchangeable with the goal. Sacred values are about how certain actions have absolute value; they cannot be traded off for material gain. The actions are not taken to serve some other end but instead because they are judged appropriate in and of themselves. Whatever consequences they produce down the road are irrelevant. Their end is doing whatever action is prescribed by the sacred value. This is a form of means-end fusion.

Extremism provides a stage to view sacred values as a type of means-end fusion. In a threatening social environment, the simplicity and absolutism of sacred values can lead to zealotry. We have seen that acts of terror are motivated by sacred values, like a terrorist's sense that detonating a bomb is an act of divine justice. Suicide bombings of Israelis by Palestinians became increasingly frequent during the second intifada (uprising) in the early 2000s. Middle East expert Mia Bloom argued that "violence has become *the* source of all honor among Palestinians. . . . Individual esteem is bound to group status, physically and symbolically. Sacrifice and risk employed on behalf of the group become valuable virtues, rewarded by social status."[7] Such actions can also be understood in terms of means-end fusion. The terrorist's intent is to wreak havoc and confusion, and that is exactly what their action involves. Their means is closely related to their ends. In this most extreme form of behavior, sacred values come to describe actions that are inseparable from their goals.

Multiattribute Utility Theory

Consequentialists have their own ways to simplify decision making. To describe them, I first need to introduce you to a theory that is more helpful than EU theory because it applies much more widely. Imagine you're a consequentialist considering taking a new job. After reading chapter 4, you might want to use EU theory to make your decision. To figure out the utility of your new job, you need to assess a lot of attributes of the job and figure out how to put them together. Jobs offer salaries and benefits, but they also offer learning opportunities and colleagues to interact with. They differ in how you spend your time, slogging over boring spreadsheets or interacting with interesting people. They are located in better or worse places and are closer or farther to where you live. They offer a greater or smaller chance for advancement. Jobs have multiple attributes, and choosing among job offers requires a strategy that takes all of those attributes into account. This is not unique to jobs. Pretty much all choices involve multiple attributes. Cups of coffee differ in size, amounts of caffeine, boldness, nuttiness, and so on. So a visit to your local barista can require integrating a slew of attributes into a final decision.

EU theory tells you how to calculate the total utility of your job given the probabilities and utilities associated with all of these attributes, but it has nothing to say about how to come up with those probabilities and utilities. In chapters 5 and 6, we talked about how to come up with probabilities. Now let's consider the problem of coming up with utilities.

Is there a best way to make multiattribute decisions? In other words, is there a normative theory of multiattribute decision making in the same way that EU theory offers a normative theory of simple decision making? The general answer is "no." The main reason is that attributes tend to depend on one another, so it is impossible to state a rule that dictates how to combine them. Making tons of money might not be my main goal when looking for a job, but if the salary a job offers is high enough, then that will take care of some other attributes. For instance, location becomes less important because I will be able to afford to live in a nicer location and pay for more expensive transportation options. I have to think about all the attributes in a holistic way to make a good decision. Attributes combine in unpredictable and idiosyncratic ways. Hence there is no theory I can appeal to in every case to tell me how to combine them.

But there is a normative theory—a theory that guarantees the best possible decision—that applies in special cases. Imagine that I have a set of, say, three attributes. For example, my job options might differ only in salary, advancement opportunities, and commute time. To make a decision, I need to put a weight on each attribute. How important is salary versus commute time? If salary matters five times as much, then I give salary a weight of 5 and commute time a weight of 1. I also need utility values on the specific attributes of each job. So if one job offers $70,000 per year and another job offers $80,000 per year, then I might give the first one a salary utility of 70 and the second one a salary utility of 80. I have to assign utilities to the other attributes on the same scale too. For instance, if a ten-minute commute is as good to me relative to other commute times as $80,000 per year is relative to other salaries, then I would assign the ten-minute commute a utility of 80. If I assigned a twenty-minute commute a utility of 70, that would imply that this ten-minute difference in commuting is worth $10,000 per year to me. By assigning weights to attributes and utility numbers to attribute values in this way, I can represent my decisions with sets of numbers.

Once I have those numbers, then aggregating them into one utility value is simple, as long as one condition is met: The attributes have to be independent. This means that their contribution to the overall utility must remain the same regardless of the value of other attributes. If I hate commuting and so a shorter commute is always better to the same extent regardless of a job's location, then commute time is independent of location. If I always want a greater amount of coffee regardless of its caffeine, boldness, or flavor, then coffee size is independent of these other attributes.

If (and only if) I have determined that all attributes are independent, then all I have to do is multiply each attribute utility by the weight of the corresponding attribute and add the resulting values for all attributes. This is called the weighted additive model (WADD). This will tell me what the total utility of the outcome is. I can then use the law of expected utility and choose the option with the highest expected utility.

Here is an example for those of you who want to work through the details. Say that I have determined that my three job attributes are independent of one another, and I have two job offers. The Widget Factory is offering me a job at $80,000 per year with few advancement opportunities and a ten-minute commute. ACME Plastics is offering me a job at $70,000 per year with many advancement opportunities and a twenty-minute commute. We decided earlier that salary gets a weight of 5 and commute time a weight of 1. Say the advancement opportunities attribute is almost as important to me as the salary, so I give it a weight of 4. We also decided that the utilities of $80,000 per year, $70,000 per year, ten-minute commute, and twenty-minute commute are 80, 70, 80, and 70, respectively. What should the utilities of few and many advancement opportunities be? I think that many opportunities should get a much higher utility value than few because it would make a big difference to my future, so I assign many opportunities a utility of 80 and few opportunities a utility of 40. Now all I have to do is multiply and add:

U(job) = weight of salary × utility of salary + weight of advancement opportunities × utility of advancement opportunities + weight of commute time × utility of commute time

Therefore U(Widget Factory job) = 5 × 80 + 4 × 40 + 1 × 80 = 640 and U(ACME Plastics job) = 5 × 70 + 4 × 80 + 1 × 70 = 740.

Even though ACME Plastics is offering me a lower salary and longer commute time from my home, it has higher utility because it is offering more advancement opportunities. That is the job I should take.

WADD has its virtues even when the attributes are not entirely independent. When trying to predict how people will behave in the future, WADD is more accurate than human intuitive judgment. Weighting and summing a bunch of objective scores turns out to be a better predictor than the judgments that experts make after a short interview, even if the experts get to see just as much data as the algorithm.[8] This has been shown in multiple domains, predicting clinical psychiatric diagnosis, graduate school successes, parole violations, case worker judgments of offender risk, spousal assaults, and more.[9] The problem with interviews is that they loom so large, we cannot ignore them. We are so responsive to human interaction that we make up our minds based on a short interaction and neglect more objective evidence gathered over a long time. The power of interviews shows the representativeness heuristic at work and illustrates how it can lead us astray. You can do better if you have objective data and are willing to do a little multiplication and addition.

In fact, to do better than human judgment, you do not even need to do the multiplication. A model that ignores the importance of attributes by just giving each attribute equal weight is also a better predictor than human intuition.[10] This is called the equal weight heuristic (EQW).

Multiattribute Strategies and Simplifying Choice

While WADD works well for maximizers with sufficient time and discipline, simpler strategies often make more sense. A classic text, *The Adaptive Decision Maker*, describes the multiple strategies that people rely on for making multiattribute decisions and how we tend to choose among them.[11] Along with WADD and EQW, there are the following:

Satisficing (SAT). SAT is a precise way of describing how a satisficer might decide. Options are considered as they are encountered. On encountering an option, there are two steps. First, compare the value of each attribute of the option to its cutoff level, the minimal acceptable level for the attribute. If all attributes meet their cutoff, you're done. Choose that option. Second, if any attribute is below its cutoff, reject the option and wait until you encounter another option. Then go to the first step. The idea is to choose the first option whose attributes satisfy all of your cutoffs. In the case that no option satisfies all of your criteria, you should relax your cutoffs and repeat the procedure. Success with SAT is partly a matter of luck. What you choose depends on the order of encountering options. If the first option you encounter happens to be a really good one, you make an excellent choice. Some lucky people meet their life partner at a young age and live happily ever after.

Lexicographic ordering (LEX). This strategy involves choosing on the basis of the single most important attribute to you. Specifically, you determine the most important attribute of the options in front of you and examine the values of all alternatives to that attribute. Choose the option with the highest value. If there is only one option with the highest value, you're done. Otherwise, determine the second most important attribute and repeat. LEX is a common strategy. It is how I did most of my grocery shopping when I was younger. I would choose the cheapest option. It is still how I do a lot of my online shopping for goods that do not matter much.

Elimination by aspects (EBA). EBA is similar to LEX except that you eliminate rather than choose. The first step is to eliminate all alternative options with values below your cutoff on the attribute most important to you. If more than one alternative remains, choose the next most important attribute and repeat. This strategy was first proposed in 1972 by Amos Tversky.[12] In his original formulation, some uncertainty was introduced to account for the fact that people don't always make the same choices. The most important attribute was chosen probabilistically.

Frequency of good and bad features (FRQ). In this strategy, cutoffs are used to separate good from bad values for each attribute. Any attribute whose value is above the cutoff is considered good, otherwise the attribute is bad. Then all one does is count the number of good and bad attributes for each alternative. Choose the alternative with the most good attributes.

Habit heuristic. This final strategy is one we have encountered before. It is the simplest of all and probably the most common. Choose what you chose last time.

Which strategy should you use? If you have enough time, we have seen that the only strategy with a strict normative justification that guarantees the outcome with the most expected utility is WADD, but only when attributes are independent of one another. So that is a good choice. Not only does it have a normative justification, it has been shown to be a good rule to predict the future. And EQW works pretty well too. The advantage of EQW is that it is simpler as you can avoid having to assign weights to dimensions. Assigning weights requires figuring out how the various attributes trade off with one another, and that can be difficult and produce a lot of internal conflict.

The Adaptive Decision Maker frames the issue as a metachoice. We have a choice to make and this leads us to a set of choice strategies. Now we have to choose a strategy. We can begin by identifying the attributes of the various strategies. The strategies differ in four key ways.

First, they are either *compensatory* or *noncompensatory*. A compensatory strategy allows attributes to trade off with one another. WADD, EQW, and FRQ are all compensatory because a really good attribute can be overwhelmed by one or more bad attributes, or vice versa. Making these compensatory trade-offs requires effort and can cause conflict. All the other strategies are noncompensatory.

Second, the strategies differ in whether or not they are selective in the information that they process. All the strategies but one ignore either some of the alternative options or attributes of the options. SAT and the habit heuristic ignore options because they allow one to choose before considering all of them. EQW ignores the relative importance of attributes. FRQ ignores the relative importance and much of the attribute information, looking only at whether attributes are above or below a threshold. LEX and EBA consider only the most important attributes. Only WADD considers all the information, and even it ignores any dependencies between attributes.

Third, only some of the strategies generate an overall utility for each outcome. SAT, LEX, EBA, and the habit heuristic all make a choice without first coming up with a utility value for options. The others do generate an overall utility (e.g., FRQ's utility is the number of positive attributes).

Finally, the strategies differ in how much and what kind of reasoning they require. WADD requires multiplication and addition, EQW only addition, and FRQ only counting. SAT, LEX, and EBA all require nothing but value comparisons. Choosing by habit requires no reasoning at all.

Each of these attributes of choice strategies imply a degree of difficulty in applying the strategy. Compensatory strategies require more cognitive effort than noncompensatory ones. The more selective the strategy is in the information it requires, the easier it is to implement. Generating a utility is harder than not generating one. And the strategies that require more reasoning demand more effort than those that require less and easier types of reasoning. So choosing a strategy involves determining how much time and effort you are willing to put into a given decision.

People are more likely to use simplifying strategies when options themselves are complex rather than simple. Choosing among hangers of different sizes does not require simplification because the choice is necessarily simple: Choose the one that fits the clothing you want to hang up. But choosing which dress to wear can be a difficult problem with many dimensions. So going with the most important attribute (LEX) or eliminating dresses that do not meet your most important criteria (EBA) makes sense. People are also more likely to simplify if there are more options. In fact, if there are enough options, one is forced to. If choosing a mate, one cannot consider all the options. That would require meeting a substantial portion of everyone alive in the world and you would be long gone before your task was completed. So SAT is pretty much necessary (with high cutoffs of course). One is more likely to simplify if under time pressure as well. If we are asked to make a decision at the checkout counter (would you like to donate to the supermarket's favorite charity?), we do not have time to do a full WADD analysis, and we are likely to resort to one-reason decision making or the habit heuristic. Stress can have the same effects as time pressure, as both make it difficult to reason in a deliberative manner.

Notice that none of the multiattribute strategies provide a way to take uncertainty into account in decision making. Uncertainty complicates choice because it means we have to consider not just a single outcome of each option but instead a range of possible outcomes, and we need a utility for each outcome. The way to incorporate uncertainty into a multiattribute decision is to integrate one of the multiattribute choice strategies with the kind of expected utility analysis we described in chapter 4. Start with the expected utility analysis: generate options, figure out each possible outcome of each option, and assign a probability to each consequence. Where you will get stuck is in assigning utilities to each consequence. That is where the multiattribute choice strategies come in. First, choose a strategy that will give you an overall utility (WADD, EQW, or FRQ) and then use it for each outcome to determine its utility.

The method is straightforward in principle. The problem is that when outcomes have a lot of attributes, there are a lot of possible outcomes. If you have merely 3 attributes with 3 possible values each, and each outcome can have any combination of attribute values, there are $3 \times 3 \times 3 = 27$ possible outcomes. And if you have 5 attributes with 5 values each, there are $5^5 = 3{,}125$ possible outcomes. The number of outcomes increases really fast with the number of attributes and attribute values. Assigning probabilities and utilities to each one becomes an art and can require some impressive mathematical and statistical concepts. Decision theorists have a lot to say about how to cope with such situations.[13]

Reasons for Choice

One question that determines people's decision strategy is whether or not they have to justify their choice. Sometimes we have to justify our choice of marketing strategy to a board of directors, or our decision to change suppliers to our boss, or our decision to quit high school and go live in a van down by the river to our mother. Having to justify a choice is made easier by some choice strategies than others. LEX makes justification particularly easy because the justification is a single reason: I chose based on the most important attribute (I quit school because I can learn so much more by fending for myself among all the vagrants down by the river). In fact, there is some powerful evidence that people do make choices this way, based on their ease of justification.

Consider this decision:

> Imagine that you serve on the jury of an only-child sole-custody case following a relatively messy divorce. The facts of the case are complicated by ambiguous economic, social, and emotional considerations, and you decide to base your decision entirely on the following few observations.
>
> **Parent A:**
>
> - average income
> - average wealth
> - average working hours
> - reasonable rapport with the child
> - relatively stable social life

Parent B:

- above-average income
- very close relationship with the child
- extremely active social life
- lots of work-related travel
- minor health problems

One group of participants in this study were in the award condition. They were asked, "To which parent would you award sole custody of the child?" Another group was in the deny condition. They were asked, "Which parent would you deny sole custody of the child?"[14]

Presumably, these questions are mirror images of one another. If you think one of the parents should be awarded sole custody, then you would probably expect that the other parent would be denied sole custody. But that is not what happened. Instead, both groups were more likely to choose parent B. Parent B was both more likely to be awarded and to be denied sole custody. Parent B can be considered the *enriched* option. That parent has several positive parenting attributes (above-average income and a close relationship with the child) and some quite negative attributes (lots of work-related travel and minor health problems). If you are looking for a reason to award sole custody, parent B provides one. And if you are looking for a reason to deny sole custody, parent B again provides one. Parent A, in contrast, is middling all the way around. That parent does not offer any really good reasons to either award or deny custody. So the preference reversal that we see in this instance could emerge because people are seeking reasons to justify their choice. The search for different reasons depending on the question asked is another example of the compatibility principle that we saw in the last chapter. In this case, compatibility with the question (award or deny?) determines the kind of evidence that people search for in order to justify their choice, and what people find (reasons for versus against) lead to opposing choices.

The need to justify in this case arises because the issue concerns how to bring up a child. It is the kind of issue that elicits a strong sense of responsibility. People want to make the right choice and be positioned to defend that choice to others, especially the child and parents themselves. The desire to justify even emerges in this hypothetical example, though, where there are no actual parents, children, lawyers, judges, or anyone else. Issues like this make us want to justify our choices to ourselves. We ourselves need a reason even when we are the ones making the choice. This need to justify to ourselves is a natural product of deliberation, of thinking about our decision. If we collaborate with others on a decision, a justification is necessary to argue for an opinion and reach group consensus. Deliberating with ourselves is not so different. In that instance, we are trying to make a case to ourselves—often with an internal dialogue—and that too requires a justification.

Justifications have to be understandable. A reason cannot be compelling if you cannot make sense of it. And a justification is better if it is articulable. That makes it easier for others to understand, and may make it easier for the justifier to understand too. Simpler ideas are easier to understand and articulate than more complex ones. In sum, thinking about a choice tends to guide us to the option that is easiest to justify, and that means using a more comprehensible and therefore simpler decision-making process. The upshot is that thinking about our decision can have the ironic effect of simplifying how we choose.

Here is a concrete example. I am on my way into an important meeting and am told I have to immediately choose a restaurant to take the client to that evening. So I scan in my mind the restaurants that I can think of. One of them ticks off the most important criteria. It is not too far, and it has a lovely ambiance, good food, and decent service, plus the company is paying so price does not matter. Decision made.

But now imagine I have fifteen minutes before the important meeting to make the decision. I have time to deliberate. I want to make sure the client is happy, that my boss is happy, and that dinner can be productive. My deliberation leads me to the conclusion that the choice I can most easily justify to both my client and boss is the restaurant that makes it easiest to have a productive yet private conversation. So ambiance is the most important dimension, and I choose the restaurant that provides the right kind of ambiance. By virtue of thinking about the problem, I simplified it and made the choice based on a lexicographic—one attribute—strategy, ignoring distance, food, service, and other attributes that I considered when my time was limited.

The availability of reasons explains other decision-making phenomena. One is called asymmetric dominance, sometimes referred to more plainly as decoy effects.[15] Let us say you cannot decide between eating an apple and an orange. I make your decision even harder by adding an item to your choice set, an apple with a worm in it. It turns out that I did not make your choice harder at all. What I actually did was to give you a reason to choose the original apple (it doesn't have a worm in it!). All else being equal, for a variety of consumer products, adding a third option that is *dominated* by one of the original two options—but not the other—increases the market share of the dominating alternative. One option dominates another option if it is at least as good as the second option on every attribute and better on at least one attribute. The first apple dominates the second apple because they are equivalent in every way except that it does not have a worm. Having that wormy apple around makes the nonwormy apple seem more attractive relative to the orange.

The asymmetric dominance effect can be explained without appealing to the availability of a reason. It could be a kind of contrast effect: The apple looks better beside an option that is similar to it but worse (an apple with a worm) than it does when placed beside an orange. Whatever the right explanation, many studies have shown that adding new options to a choice set can actually increase the market share of old items.[16] This violates a common assumption of decision theory called independence of irrelevant alternatives—that if A is preferred to B, then making C available should not cause B to be preferred to A.

These considerations have been amalgamated into a theory of judgment and decision making called query theory that posits that people make judgments and choices by asking themselves a series of mental questions or queries.[17] Each question is answered in order. In query theory, questions that are answered earlier in the sequence have more influence because people only have a limited amount of time and effort to spend on any given judgment or choice. Moreover, earlier answers can interfere with later ones by framing them. To see how it works, consider the endowment effect, the finding that people demand more money for an object they already own than they are willing to pay for the same object if they do not own it. According to query theory, the endowment effect occurs because the two situations elicit different questions. If you already own the object and are asked to sell, the first question you might ask yourself is, *What reasons are there for owning this product?* In contrast, if you do not own it and are considering buying, the first question you might ask yourself is, *What reasons are there for not owning this product?* Coming up with a good answer to the first question provides a reason for owning the product. This inflates its perceived value.

A good answer to the second question provides a reason not to own it and this deflates its perceived value. Hence the product becomes worth more to someone who already owns it.

Choosing with Stories

Choosing with reasons does not require that we focus on only a single reason. Here is how polymath Ben Franklin described his own decision-making process in 1772:

> My Way is, to divide half a Sheet of Paper by a Line into two Columns, writing over the one *Pro*, and over the other *Con*. Then during three or four Days Consideration I put down under the different Heads short Hints of the different Motives. . . . When I have thus got them all together in one View, I endeavour to estimate their respective Weights . . . [and] find at length where the Ballance lies. . . . [When] the whole lies before me, I think I can judge better, and am less likely to make a rash Step; and in fact I have found great Advantage for this kind of Equation, in what may be called *Moral* or *Prudential Algebra*.[18]

Franklin listed as many reasons as he could so that he could see the problem as a whole. He wanted to see all the pieces together in one place before he assembled the item.

What did Franklin mean by "when the whole matter lies before me"? What does it mean to see the problem as a whole and put a set of reasons into a coherent package? It means to have an all-embracing mental representation of relevant material, an idea in one's mind that makes sense of all the reasons at the same time. The form that such a representation takes will depend on the nature of the decision. If you are choosing among designs for a house, then the all-embracing representation might be a mental blueprint for the house. If you are choosing among recipes, it might be a mental image of the dishes' appearances and taste. If you are choosing among ways to go about fixing a broken appliance, the mental representation might be a causal model that describes how the appliance works.

Often our choices depend on beliefs about how agents behave. Choosing among job applicants is a choice about who will be most diligent, reliable, and productive. Deciding whether an accused is innocent or guilty is about understanding their motivations and intentions as well as their actions and the consequences of those actions. These kinds of considerations are generally represented by people in the form of a story.[19] Some evidence that people represent information in the form of a story comes from the finding that a more effective way to persuade somebody is to offer them a relevant anecdote rather than an abstract argument, even though people believe they are not taken in by stories and are more responsive to logic.[20]

Psychologists Nancy Pennington and Reid Hastie offer more evidence for what they call the story model of decision making. They asked some students to imagine they were jurors in a court case. The researchers gave the students evidence about a crime, and the students' job was to decide if the accused perpetrator was guilty or not. They got the evidence in one of two forms. Either it was shown to them in the order reported by witnesses, as follows:

The first witness is a police officer who states

- that he saw Johnson and Caldwell outside the bar
- Johnson laughed at Caldwell
- Caldwell punched Johnson in the face
- Johnson stabbed Caldwell in the chest

The second witness is a medical examiner who reports that

- Caldwell died from a knife wound in the heart
- Caldwell was carrying a razor
- The weapon was a sharp kitchen type knife
- Caldwell was a big man, weighing about 220 pounds

The third witness saw

- the defendant Johnson, and the victim, Alan Caldwell, quarreling early that day

And then students were asked whether or not Johnson is guilty of murder.

A second group got the same evidence, but this time in the form of a story. It was chronological and provided causal links between episodes, clarifying actors' intentions:

> The defendant Johnson, and the victim, Alan Caldwell, had a quarrel early on the day of Caldwell's death. At that time, according to evidence from the medical examiner, Caldwell threatened Johnson with a razor. Caldwell was a big man, weighing about 220 pounds. Later in the evening, they were again at the same bar. A police officer reports seeing Caldwell punch Johnson in the face, then Johnson laughed at Caldwell before stabbing him in the chest. Caldwell died from a knife wound.[21]

Subjects in this story-order condition were more likely to convict Johnson of murder. Even though the evidence they saw was essentially identical, reading the story left a stronger impression of the accused's guilt than seeing the evidence in a more arbitrary order. Pennington and Hastie argue that people are not driven by evidence per se but rather construct stories to make sense of the evidence when making decisions in law, medicine, politics, diplomacy, and elsewhere.[22] These stories can be chock-full of norms and assumptions as they require us to fill in many gaps. For instance, why would Johnson laugh at Caldwell? The inference that it was a show of bravado is one you might have made based on your understanding of US culture.

Stories provide a way to bring coherence to data that would otherwise be hard to comprehend. Stories have an underlying causal structure that explains how episodes relate to one another and provide motivations for characters' actions, sometimes in the form of sacred values. Traditionally, stories are presented chronologically, allowing events to be encoded according to a meaningful timeline that makes it easier to construct an explanation for events.

The Value of Simplicity

We have seen that not everyone aims to simplify their decisions. There are maximizers out there who fiendishly collect information and devote lots of resources, cognitive and otherwise, to finding the best option. I suspect that even such people do not always maximize but instead sometimes satisfice. They might maximize their shopping but satisfice when it comes to what music they listen to. People are multidimensional, and there are not enough hours in a lifetime to maximize all the thousands of daily decisions we make.

Most of us, however, know better than to devote too much of our lives to decision making and so we simplify our decisions, even when we are aware this might reduce the quality of our decision. We simplify in a variety of ways. We have rules to live by that let us bypass some decisions. Some of those rules tell us to be creatures of habit, and others encode sacred values that allow us to bypass difficult moral and social dilemmas. Such rules can become so ingrained that they become intrinsic motivators; our actions become so fused with their objectives that we no longer distinguish our behaviors from our goals. This can be dangerous when our behavior and goals involve imposing our will on people who are resisting.

We also simplify by deploying choice strategies that guide us to ignore some of the aspects of decisions that are difficult, like how to trade off attributes, deal with complex options, and wade through massive numbers of options. Different decisions call for different choice strategies, each of them simplifying in different ways.

When thinking about simplification, we should remember one of the lessons of chapter 6. Our more important decisions are made in the context of other people. Either we are deciding in conjunction with others, or we are heavily influenced by others, or our decisions have implications for others. This is why reasons turn out to be such central elements of so much decision making. We like our decisions to feel justified. We want to be able to convince others and ourselves, or at least explain why we chose what we did. We might even want the reason for our decisions to have a lasting impact by influencing future decisions in the same way that court rulings set a precedent that can influence later court rulings. This can only happen if we have a reason for our choice.

Because we live in a world framed by time and causality, populated by people and other animals loaded with agency and interests, we use narrative to make sense of our experience and generate reasons. We tell ourselves stories to bring coherence to complicated events in order to understand them. Simplicity is nothing if not making events understandable, so these stories allow us to simplify our complicated worlds. Unfortunately, living with stories forces a kind of departure from reality. The tale I tell about, say, our evening together will not be exactly the same as the tale you tell. When groups latch on to different narratives and those narratives come to clash, our decision making can lead to catastrophe.

But our decision making often leads in a much more comforting direction as we strive for our own comfort, health, and joy as well as that of others. Having a machine that made all of our decisions for us would simplify and sometimes be desirable. Yet we would lose the spark that makes us human and the opportunity to make the calls that give our life meaning.

9 How Should I Choose? Consequentialism versus Sacred Values

We have now described very different strategies to make a decision. One strategy focuses on how to get the best consequences. The other strategy focuses on which action is most appropriate. You might think that the distinction is between making decisions to maximize usefulness—that is, achieving the best outcomes—versus making decisions to be most righteous—namely, taking the right action. Maybe consequentialism is just a way of being pragmatic, or instructions for how to accomplish tasks you are faced with. In contrast, sacred values provide a way to assess what is moral, to determine what you should do to be a good person, regardless of the task in front of you.

On this interpretation, the distinction may seem retrogressive, cutting against some of the most celebrated scientific and philosophical insights of the nineteenth century. After all, the greatest scientist of the era, Charles Darwin, argued that altruism could be explained in terms of the fitness of groups, as an attempt to reduce one aspect of morality to behavior that is useful (to a group). Analogously, utilitarians like John Stuart Mill contended that morality could be reduced to costs and benefits. According to both of them, issues of right are, deep down, issues of usefulness. Perhaps the distinction between consequentialism and sacred values should be heaped on the dustbin of history with other bad ideas like phrenology and phlogiston theory. If this is what you think, not everyone agrees with you. Even in the nineteenth century, one scientist, George Jackson Mivart, complained that evolutionary biologists and utilitarians did not appreciate the "difference between the ideas 'useful' and 'right.'"[1]

I am not so sure that the contrast between useful and right does characterize the distinction between consequentialism and sacred values. First, consequentialism may not only be about usefulness; it might also be about what is right. It could stand as a moral theory (and is often treated that way).[2] Who is to say that utilitarians are not correct when they assert that the most moral behavior is the behavior that is most useful in the sense that it promotes the most benefits and avoids the most costs? Thoughtful utilitarians such as Mill were careful to argue that what matters is the benefits and costs that everyone experiences; that utility is not selfish. What could be more right than to behave in a way that maximizes everyone's happiness? On top of that, sacred values are not necessarily about what is right. Some sacred values are about how people of various castes or communities should dress, or about religious observances. It is hard to maintain that these behaviors are moral to someone who has not been socialized already to see them that way. They seem to be more about ritual and social role—about what is useful—than about morality. Indeed, guidance by sacred values may well be useful. As we have seen, good decision making requires simplification. There are not enough hours in the course of a lifetime to deliberate over every decision. We must simplify. One way to simplify is to have rules about which actions to take. Sacred values supply such rules and also serve other important functions, like helping to tie a person to their community.

What does distinguish consequentialism and sacred values is that they refer to different frames of reference. How one approaches a decision depends on the perspective of the decision maker in the same sense that how one sees a fire depends on perspective. It can be seen as a source of warmth or, alternatively, a source of carbon emissions. Some choices are easier to frame consequentially. The decision to buy a tube of toothpaste is more naturally framed in terms of consequences than in terms of sacred values. Other choices are more naturally viewed in terms of sacred values, like the decision to recite an oath that everyone around you is reciting. But even these choices are matters of framing. I might have a sacred value about not supporting large corporations that make toothpaste and be stuck brushing my teeth with soap. And I could take a consequentialist position on reciting the oath by making the decision based on what I will communicate to the person beside me.

The frames differ in their focus—consequentialism on consequences and sacred values on actions. And they differ in their absolutism. Consequentialism allows trade-offs with other consequences and sacred values do not (in principle). But any discussion of decision-making strategies requires that we respect the distinction between normative and descriptive theory. The normative theory of sacred values does not allow taboo trade-offs, doing the wrong thing for the purpose of material gain. Descriptively speaking, though, people do have their price. A tempting offer can cause someone to switch from a sacred values perspective to a consequentialist one. I might be willing to break a social taboo like wearing a fig leaf to a yoga class if, for instance, someone pays me enough to do so. I can then do a lot of good for other people with the money.

The normative/descriptive distinction applies to consequentialism too. I might take a consequentialist perspective for the normative reason that I want to maximize happiness, but the preceding several chapters have documented a range of situations that lead people to fail to maximize even when they are trying to. If I am aiming to go on the most pleasurable vacation, my choice process might get the probability of a pleasant hotel wrong, or oversimplify and ignore a critical attribute like whether or not the hotel has air-conditioning during a heat wave. If that happens, my consequentialist decision process could go horribly awry.

A False Dichotomy?

Psychology is full of dichotomies, such as nature/nurture, introvert/extrovert, or male/female. In each case, these turn out to be oversimplifications and the truth requires more nuance. Is that also true of the consequentialism/sacred values dichotomy?

Normatively speaking, I would vehemently defend the distinction. Consequentialist justifications for decisions are different animals than sacred value justifications. Consequentialist justifications appeal to causal mechanisms, probability, and cost-benefit analysis. They are arguments about how to achieve the best outcomes. A consequentialist justification for a vote for a particular political candidate will reference the politician's policy positions and how likely they are to be realized. In contrast, sacred value justifications require logical derivations from fundamental religious or philosophical presuppositions. They make use of deontological reasoning. They are arguments about which action is most appropriate. A sacred values justification for a vote would reference the politician's values and explain why you agree with them. The two kinds of justifications make different assumptions, use different kinds of logic, and have different goals. They have little in common.

Descriptively speaking, the issue is more complicated. How do people construct the two kinds of justifications? Joshua Greene argues for a dual-process model of moral judgment, where one process is more analytic and corresponds to consequentialist reasoning, and the other is more emotional and intuitive and corresponds to deontological responding.[3] While I agree that we have distinct cognitive systems underlying deliberate, analytic versus automatic, intuitive reasoning, I do not believe those systems are specialized for consequentialist versus sacred values reasoning.[4] Both kinds of responses can involve deliberation. Complex reasoning about the consequences of, say, foreign policy options requires deliberation, but so does complex reasoning about sacred values involving the rights of the unborn or prison inmates. And both kinds of responses are sometimes intuitive. The decision to avoid an obstacle on the highway is made automatically, presumably because of the negative consequences of not avoiding it. And the sacred value that prevents most people from harassing others that they are attracted to is highly intuitive too. In the trolley problems that we will encounter next, it appears that consequentialist responding involves more deliberation than nonconsequentialist responding. But Greene's dual-process model claims that only consequentialist reasoning should require deliberation and it always should. That is a much stronger claim without empirical support.

If the two kinds of decision making are not governed by different cognitive systems, then what does it mean to say that consequentialist and sacred values provide different frames for the reasoning involved in decision making? Perhaps the two kinds of reasoning involve overlapping cognitive mechanisms that are not cleanly distinguishable? Surely both involve processes of language comprehension, abstract reasoning, comparison operations, and evaluation? In that sense, perhaps there is no real distinction; rather, each kind of reasoning draws on a set of common resources.

Another way to state the concern is to point out that there are multiple tools people have to make decisions. We can determine outcome values, calculate expected utilities, and apply religious values, moral values, conventions, legal dictates, and habits. All of these involve rules in one or another sense, and each of them has, to a greater or lesser extent, an emotional underpinning. Why pick out consequentialist and sacred value reasoning? Are they not just two out of a large number of ways of slicing the pie of decision-making methods?

It is true that the two kinds of processing must rely on common cognitive resources, but this does not mean there is not a real, substantive distinction between them. The psychological reality of each of these decision procedures is grounded in the fact that decision makers demand that their decision strategies make sense. For most of us, making decisions via a coin flip is a last resort. Why? Because, as we have seen, people are not satisfied with a decision unless they can justify it. This is especially important when our decisions are social—that is, when we are making them with others, if our decision impacts others, or if we have to justify ourselves to others. But that is not necessary; we want our decisions to be justified even when other people are not involved in the equation. Even alone, we need to feel that our choice is vindicated.

Consequentialist and sacred values reasoning both offer their own justifications for our decisions. There are other types of justifications as well. We might choose an option that brings us relief from pain or anxiety. We might choose to smoke a cigarette or engage in some other harmful behavior on this basis. But such a reason is more an excuse than a justification. It lets us know why we decided as we did but it does not make us feel satisfied that we made the right decision. It just leaves us wishing we had the willpower to make the decision on a different basis.

People strive to do better than find respite from negative feelings. We aspire to make decisions that we are proud of because they serve our long-term ends, are consistent with our values, or both. We want our decisions to be impervious to other people's objections, to have an unassailable logic. The only kind of decision procedures that achieve that are ones that are consistent with a normative theory. Normative theories are, by definition, based on unassailable logic (or as close to unassailable as we can get). To feel like we have made the right decision, our descriptive theory of decision making must converge with a valid normative theory. This is more obvious when our decisions involve other people. Then we really have no choice but to rely on a decision procedure with a compelling normative justification. Yet it is true even when others are not involved. Descriptive and normative theories are not the same. Only normative theories provide a compelling rationale. Hence normative theories become descriptive when we justify ourselves effectively.

The fact that consequentialism and sacred values provide different and compelling normative justifications means that they are also different and compelling bases for describing choice. People use them. And the two kinds of reasoning offer different rationales and thus represent a real dichotomy, not a false one. A person might make a compelling decision about how to spend their Sunday morning by maximizing gains and minimizing losses. Someone else might make the decision by appealing to a sacred value embedded in a religious practice. Both serve as compelling justifications, but they appeal to different people. The correct decision procedure to use is the one that the people you care about will agree with. Hopefully, you yourself are someone you care about.

Blissful Ignorance

None of this is to say that people are conscious of all of this. Indeed, most people generally are blissfully unaware of the difference between types of normative justifications. Most of us will use every argument at our disposal to make a point. We will appeal to consequences or sacred values, and we will often not distinguish between them. Consider an argument like, "The sanctity of marriage would be undermined by allowing gay people to marry because it would destabilize society." The argument uses a consequence (the destabilization of society) as a reason in favor of a sacred value (maintaining that marriage must be between a man and a woman). If the consequence were true (it is not), it might argue against allowing gays to marry. But it does not support any claims about the sanctity of marriage. The justification of a sacred value comes from the correctness of the action it refers to, not from its future consequences. Just as a sacred value should not be violated for the sake of material consequences, it should not be bolstered by them.

The conflation of consequences and sacred values is par for the course when people are constructing arguments. In work I did with Babak Hematian on South Korean attitudes toward unification with North Korea, we found that those South Koreans with stronger sacred values also felt able to make better consequentialist assertions.[5] And those who claimed to be more driven by their morals and values had more faith in their own consequentialist analyses; they thought they understood the situation and issues around unification better. People do not make clean distinctions between consequentialist analyses and sacred values. This is most obvious when people are trying to persuade others. They will justify their position however they can, with logic and history, by appeals to morality, by making relevant players heroes or assassinating their characters, or by any other means at their disposal.

The very word "belief" hints at the confusion. I have been using belief to refer to one aspect of consequentialist reasoning—our mental representations about how the world works. Specifically, how decision options lead to outcomes. But in natural language, beliefs can also be about sacred values. I might believe that a woman has a fundamental right to choose or that eating meat is evil. Language does not help us to keep the sacred values/consequentialist distinction straight because human beings often ignore it.

The distinction between consequences and sacred values is just not top of mind most of the time. People can recognize an appeal to their interests as opposed to an appeal to their conscience, but we will throw the kitchen sink at an argument, even confusing our beliefs and values. As a result, people with different interests will frequently have different beliefs that provide support for those interests. A greater percentage of women than men agreed with the statement, "There is a difference between the wages for women who work full time and for men who work full time."[6] Those for whom the pay gap matters more believe that it is more likely to exist too. People's interests are one source of their values, and it turns out that their values affect their beliefs. There is a good reason for this interdependence between values and beliefs. We tend to harness all of our resources to support the conclusions that we are motivated to arrive at.

Reductionism

Another reason that people might consider the distinction false is because sacred values can be reduced to consequences, and the importance of consequences can be reduced to sacred values. At some level, everything is consequentialist. Sacred values may be grounded in consequences. Some have argued that all of our political positions are grounded in self-interest.[7] Evolutionary theorists who want to ground all behavior and cognition in processes of natural selection are likely to agree with such a statement. After all, natural selection operates by selectively reproducing those traits that produce better outcomes—outcomes that are more likely to lead to the survival of an organism's genes. Sacred values that state that abortion is wrong may have evolved to facilitate the reproduction of a group's genes in societies with a high infant mortality rate, thus ensuring more children. Sacred values that express a woman's right to choose might have evolved in societies with limited resources, so that being more selective of the time and place for keeping an infant was a more effective strategy for genetic reproduction. These are merely just-so stories that I made up. But evolutionary stories like this can be told about all sacred values. Sacred values might well have their origins in the outcomes they produce.

On the flip side, consequentialism itself depends on sacred values. To be a tried-and-true consequentialist, one must be a sort of scientist. One must take data seriously in order to state with any confidence which outcomes will occur with what probability. Such a view takes empiricism to be a sacred value. More fundamentally, a consequentialist is likely to take consequentialism seriously. They are likely to hold that the best way to make decisions is by maximizing benefits and minimizing costs. This too can be construed as a sacred value.

So sacred values can be reduced to consequences and consequentialism to sacred values. Does this imply there is no distinction to be made between them? Of course not. The question of this book is how to make decisions. Any specific decision at a given point in time will be made using a particular decision procedure. Consequentialism and sacred values provide different decision procedures that each come with a solid justification. The questions of whether sacred values' origin is in consequences and whether consequentialism implies certain specific sacred values are worthy of study, but irrelevant to the question of whether the two decision procedures differ.

Trolleyology

A moral philosopher, Philippa Foot, posed a hypothetical moral dilemma in 1967 that turned into a tidal wave of psychological research in the 2000s. In a discussion of some difficult problems in moral philosophy having to do with abortion, Foot considered the case of a "driver of a runaway tram which he can only steer from one narrow track on to another; five men are working on one track and one man on the other; anyone on the track he enters is bound to be killed."[8] Foot takes it as a given that the driver should steer toward the track with only one worker, thereby saving five. She contrasts this with the case of a mob about to rain vengeance on a community unless a perfectly innocent individual is put to death. Foot takes it to be much less clear what to do in this case. Why is it acceptable to kill one to save five in the tram case but not among the rioters? The answer discussed extensively in the philosophical literature concerns the doctrine of double effect. The doctrine states that an action that has a morally positive and morally negative effect is permissible as long as the negative side effect is not intended. In the tram case, killing the one worker is not intended. It is an unfortunate side effect of steering the tram away from the five workers. Yet in the case of the riot, the community is asked to intentionally kill a perfectly innocent person, so the doctrine does not apply.

Foot's tram problem morphed into a trolley problem as it moved from England to North America, and it became a focal point for investigations of how people make moral judgments as it moved from the halls of philosophy to the laboratories of psychology. The seminal work was done by Greene and his colleagues. It is this work that motivated their dual-process model of moral judgment.[9] One process is conscious and deliberative and makes utilitarian judgments; the other responds emotionally and drives intuitive responses that are associated with deontological judgment (in this case, refusing to kill one even to save several others).

In their work, they pose a trolley problem—similar to Foot's scenario—and contrast it to a footbridge dilemma. In the footbridge dilemma, a trolley is rolling down a track about to run into and kill five workers. Before it gets to those workers, however, it must pass under a bridge. You happen to be standing on the bridge beside a fat man. You can push the fat man over the bridge to stop the trolley, saving the five workers, but in the process kill the fat man. The assumption here is that you are not fat enough yourself to stop the trolley, so it would be useless for you to sacrifice yourself. The overwhelming majority of people consider it immoral to push the fat man off the bridge even though they are willing to steer the trolley onto a track to kill one, thus saving five. This difference between the two scenarios appears to be a cultural universal. Some Asians tend to be less utilitarian than Westerners, but still show the difference between the original trolley and footbridge problems. The difference holds in a large number of cultures spanning the globe.[10]

The two different responses observed in these dilemmas correspond to consequentialist reasoning, on the one hand, and sacred values reasoning, on the other. The preference to kill one over five is a preference for one outcome over another. An unwillingness to kill one even to save five can only stem from a refusal to perform a certain kind of action, killing. These two kinds of responses have now been explored in some depth. The evidence indicates that they differ in the conditions that elicit them and the brain areas—and therefore the type of information processing—that mediate them.

Greene and his colleagues did not merely give their participants the moral dilemmas to judge; they did so while they scanned their brains in an fMRI machine. They found that certain parts of the brain lit up while giving the response that it is okay to kill one to save five—a part of the brain associated with abstract kinds of deliberative reasoning. The need for abstract reasoning to come to the consequentialist response explains another finding. People are less likely to give the consequentialist response when they are put under cognitive load, forced to think about something else at the same time and therefore unable to focus on just the trolley problem.[11] Consequentialist reasoning on this problem requires the mental space to do abstract reasoning.

A different part of the brain became active while asserting that killing even one person is not okay. This second brain region has been associated with more affective responding.[12] Moreover, the sacred value response produces a stronger skin conductance response, generally taken as evidence of activity by the autonomic nervous system or involuntary somatic reactions.[13] The sacred value response appears to be mediated by a physical and more emotional reaction, and less by abstract reasoning.

There are different kinds of emotional reactions that increase the likelihood of eliciting a sacred value in response to a trolley dilemma. One is a reaction to the potential outcome. One might become upset by thinking about the suffering of the victim who will be sacrificed. This kind of aversion to a terrible consequence makes it hard to kill someone regardless of the number of lives you would be saving in the process. This is different from being averse to the action of killing. Making an action seem more painful also leads to less utilitarian responding, but it involves a different kind of emotional response. Action aversion is more likely when the action is a direct result of the agent using their muscles to impart a force. This is why pushing the fat man off the bridge is so distasteful compared to pulling a lever.[14] It involves an intimate kind of personal force that the agent imposes through physical contact with the victim. The most developed framework for conceptualizing action aversion claims that social animals respond to distress cues in others, like fearful and sad facial expressions, body postures, and language intonation.[15] Feelings of empathy make these cues aversive and lead animals to avoid the behaviors that cause them. Over time, animals become conditioned so that merely thinking about performing actions that harm others becomes aversive and such actions are suppressed.

Both outcome and action aversion elicit a somatic response that leads to a sacred values frame. In contrast, people who do not have strong aversive reactions to others' distress are more consequentialist. This is the usual explanation for why psychopaths and those with psychopathic tendencies are more likely to kill one to save five.[16] A point made by Guy Kahane and his colleagues that I discussed earlier is of particular relevance here. Psychopaths are not utilitarian in every sense.[17] They may be more willing than nonpsychopaths to sacrifice others for the greater good, but there is no evidence that they are more willing to sacrifice themselves for the sake of others.

A final point regarding trolley problems: People have strong intuitions about what is the moral response when confronted with these problems. But that does not mean that they would actually take the action that they consider correct. A recent made-for-TV study asked whether people would steer a trolley (well, actually, a speeding train in this case) onto a track that would kill one but save five when they were dealing with a real train and real workers who (subjects believed) would actually die. The methodology used was very convincing, although the study did not test enough subjects for the results to be fully persuasive. What it found was that most people will not actually take the action required to steer the train onto a different track that would kill one. Most people just froze at the moment of decision, unable to convince themselves to take any action at all.[18] Perhaps the force of the sacred values frame was too strong for them to overcome it even when they thought they should.

How to Decide

What are the lessons of this work for how to decide? I am not a philosopher of ethics, and my role is not to tell you what the right moral theory is. The best I can do is to try to draw some conclusions about which decision-making process will leave you feeling like you did the right thing in the long term so that you will be comfortable with your actions in retrospect. This will not be of help if the decision involves sacrificing yourself because such a decision might leave you without a long term. But hopefully our conclusions will apply to such decisions too.

Most people have strong intuitions encouraging them to both achieve the best consequences and obey their sacred values. That is why trolley problems are a dilemma and so many people freeze when forced to choose between the two choice principles. Thus the first lesson is to know thyself. What do you really care about in the long term? Do you aspire to be the kind of person who deliberates carefully about consequences and maximizes, or the kind of person who sticks closely to moral principles encoded in your sacred values? People clearly differ on this dimension.

What makes identifying who you are difficult is that you may not always be the same person. Different situations elicit different frames of reference, and we cannot simply ignore that. Flipping a switch is not the same as pushing someone to their death even if the final outcomes for other people are the same. Maximizing everyone's benefits is a great justification for action, but you need to take seriously any sacred values that are calling out. It would be hard to live with the knowledge that you pushed someone to their death or allowed a crowd to lynch a person that you know to be perfectly innocent. This is exactly the kind of dilemma that brings the consequentialist versus sacred value dichotomy to life. Sometimes getting the best consequences requires doing something downright evil, that is, taking an action that violates a sacred value.

British spies were faced with just this kind of dilemma in World War II.[19] Nazis were aiming their rockets at London, and Nazi spies were sending back information about their accuracy so they could improve their targeting. But the spies were British double agents working with MI5, the British Security Service. They realized they could save twelve thousand casualties a month by misleading the Germans, prompting them to send their rockets into the less densely inhabited outskirts of London. On learning of MI5's tactic, Herbert Morrison, a minister and member of the British war cabinet, objected that the spy agency had no right to target British citizens. This violated the sacred value of do no harm. MI5 ignored the concern and continued to mislead Germans until the end of the war. The Security Service decided that minimizing casualties was more important than violating a sacred value.

Fortunately, few of us are faced with such stark moral dilemmas very often. But we sometimes are. As Peter Singer emphasizes, we could be choosing between doing a cost-benefit analysis and violating a sacred value every time a worthy charity asks for a donation. Medical professionals and charities that battle starvation have to make such choices when they distribute limited resources to a vulnerable population.

One cannot argue with making decisions to always maximize benefits and minimize costs. In the long term, always making decisions that maximize expected utility will, by definition, leave us feeling like we could not have done better. Yet given all that we know about the heuristics people use and the simplifying assumptions people rely on to make consequential decisions along with the systematic errors that result, I would not jump to the conclusion that we should always be guided by consequences. We make a lot of bad calls. One alternative is to try to correct for these errors in order to do better. That is difficult.

What is the alternative? If everybody behaved in a way consistent with broadly accepted sacred values like do not kill, I suspect the consequences to society would be pretty good. On top of that, we would all feel more comfortable with our actions. The problem is that many sacred values are not broadly accepted but rather are matters of dispute—often intense dispute. Is abortion acceptable? What about euthanasia? Do various peoples (Kashmiris, Africans, Native Americans, or Palestinians) have sacred rights to certain parcels of land (in Pakistan, Zimbabwe, the Americas, or parts of Israel)? Do women have the right to an education? Each of these questions has fanatic proponents on each side (you are probably one of them). Choosing by sacred values makes sense in principle and is certainly endorsed by members of the decision maker's own tribe. But it is not only hard to justify to those who do not share the sacred value, it usually just makes them angry. If reducing conflict is one of your sacred values, you probably should try to avoid choosing by sacred values.

This all comes out as an argument for consequentialism. Nevertheless, avoiding sacred values is harder than it appears. In chapter 2, we first saw how important sacred values are to representing one's social identity. If a member of your tribe is having a dispute with a member of a different tribe, it takes an enormous strength of character and political capital to choose the other team's side. If two sides are fighting over who owns an apple tree, it takes a powerful person to move against the tide and support the other side. Human beings almost always side with the people they are surrounded by. After all, you have to live with them. Supporting them in conflict is a prerequisite for maintaining goodwill.

What if the conflict is within your tribe? What if, for instance, your siblings are squabbling over the last piece of cake? Behavioral scientists Peter DeScioli and Robert Kurzban have studied how bystanders choose sides when others are engaged in conflict. They contrast three ideas. One idea is that people will side with whoever has higher status. This has the advantage of aligning yourself with those with more power but it also incentivizes higher-status people to exploit others. They will always get what they want if everyone supports them. Take this strategy and your big brother will always get the cake. The second idea is that bystanders choose sides based on whoever they are already closer to. The rule would be to side with the person you already have a relationship with. This is a version of the choose-your-own-tribe strategy, it just takes place within a tribe. The problem here is that it leads to gang warfare as bystanders choose different sides. This strategy guarantees conflict as the siblings will divide up into affinity groups over every cake. DeScioli and Kurzban argue for a third idea that avoids the costs of aligning with the powerful or your own subtribe. They suggest that people rely on the morality of the disputants' actions in order to decide what to do. By choosing sides based on an assessment of the morality of an action, bystanders will all choose the same side as long as they have a common view of morality, and this will encourage people to behave morally while avoiding hostility and violence. If one sibling argues for the cake based on fairness and the other based on power, then the first sibling has made a proposal with a stronger moral foundation and deserves your support.

This idea that we rely on the morality of actions is not unlike relying on sacred values. Sacred values concern action. So another theory closely related to DeScioli and Kurzban's is that people choose sides in disputes within their communities by appealing to disputants' sacred values—the values that are guiding the actions that led to the dispute. We side with whoever's sacred values align better with our own. Most of the time, this means siding with whoever's actions are most consistent with the group's values. In that way, sacred values provide the cues that people use to form alliances and moderate conflict. We can have the sacred value that everyone deserves cake or that big brothers deserve more cake. You can align with the view that suits your fancy.

There is even evidence that people prefer social partners who make judgments based on sacred values rather than consequences. People tend to believe that those who are unwilling to throw the fat man off the bridge in the trolley dilemma are more trustworthy than someone who is willing to kill one to save five.[20] Who wants to put their trust in a die-hard consequentialist? Basing your judgments on sacred values is a good strategy for getting along with others.

And that is what we actually do when evaluating other people, at least in comparison to how we judge machines.[21] Imagine you are told about an excavator who accidentally dug up a grave. Would you forgive such an action? It turns out that people's reactions to the event depend on whether the excavator is a human or machine. People are more likely to forgive the action if it was performed by a human than by a machine. The reason for this seems to be that when judging humans, judges focus on their motivations and intentions. Was the action carried out for good reasons? Was the person intending to benefit others? If there is the possibility of answering these questions affirmatively, we allow people a margin of error. In other words, we are willing to forgive if we can ascribe the person a favorable sacred value that motivated the action. But when evaluating machines, what matters is the outcome. Rather than try to figure out the machine's motivations and intentions, we judge the machine by its results. Our judgments of machines are more consequentialist, although less so when we perceive the machines as having agency and humanlike experience.

People's sacred values are indicative of their social identities and therefore characters. And we are naturally responsive to cues to those values. I would never recommend fighting these natural instincts. Still, we do need to remain aware how this responsiveness can lead to conflict because others have different sacred values, and sometimes competing ones. When your colleagues or neighbors are talking up the importance of reducing taxes, or increasing diversity and inclusion, you are learning important aspects of who they identify with and what they think. But before making a decision about their characters or whether or not to support their cause based on whether or not you resonate with the sacred value that you are attributing to them, keep in mind that decisions have consequences. If you fail to consider those consequences, you might end up acting in a way that leads to a world you do not want to live in. Reducing taxes too much can lead to roads that are full of potholes along with a country that cannot defend itself or feed poor children. And pursuing the goal of diversity and inclusion single-mindedly can leave behind a lot of deserving, hardworking people from the majority culture (possibly making them angry enough to cause a lot of trouble). Consequences matter. At the same time, our ability to correctly determine consequences is limited. So maintain some humility about what you think is going to happen, and show respect for your own and others' sacred values.

10 What About Decisions That Don't Involve Thinking?

Human beings have a habit of fooling ourselves into thinking that we are in control of our decisions, and that we consciously make decisions for reasons that we are aware of and can articulate. We may not be proud of those reasons. We may admit that we acted out of jealousy or anger. But we tend to think we know why we acted. I think I chose the pink carnation because it is my favorite color or because I operated according to a rule I hold dear—always choose pink.

Yet too often the reason for a choice only emerges after the choice has been made. I might discover what I like by virtue of what I have chosen. I keep choosing pink, so I conclude that I love pink. Whether or not I am aware of my preference for pink at the moment of choice, my reaction to pink is in fact the cause of my choice. In one case, it is a conscious preference, and in the other case, it is affecting me unconsciously; I only discover its power after I have chosen. Either way, people tend to believe they acted consciously, as if they knew all along what drove their choice. This offers an effective way to justify questionable behavior while you are doing it ("I bought the big bottle of vodka because I love the design of the bottle and can use it as a flower vase"). And if you don't know what was driving your choice, then you can make up a reason after the fact ("I kicked the dog because it is so disobedient," ignoring the fact that you were passed over for a promotion earlier that day).

Whether a decision was made consciously or unconsciously, even if the same reasons apply, matters in several ways. Unconscious decision making makes for plausible deniability to oneself. If I hire someone mainly because they are attractive but I do so unconsciously, then I can deny that attractiveness was the reason to my boss, colleagues, spouse, and myself while maintaining faith in my own honesty. Unconscious decision making affords a kind of self-deception. It also provides a legal shield. If I hire someone because of the color of their skin but I am not aware of having done so, then I will be more convincing in court. Unconscious decision making is also more malleable. If someone does not know the reason for their decisions, they may be easier to influence. Say I like a certain fast-food item because it is full of sugar, but I don't consciously know it. The company selling it can tell me that it's low sodium, convincing me it's a healthy choice. Advertisers attempt to plant reasons for us to prefer what they are selling all the time. It takes a fair amount of awareness to overcome this kind of sinister manipulation.

So far, this book has assumed that we tend to be aware of the reasons for our choices or at least that decision making involves a cognitive process that we can make ourselves aware of. The sacred value/consequentialist distinction is very cognitive. It describes two types of thought, one about action and the other about outcomes. The distinction may seem irrelevant if human decision making is largely controlled by psychological and physical forces under the hood—forces that we are unaware of or have no control over, or both. This chapter will review some key findings demonstrating the importance of unconscious processes in decision making. It will show that even unconscious decision making is a communal process, and that this makes the distinction between sacred values and consequentialism as relevant as ever.

The Vast Majority of Processing Is Unconscious (and Fast)

Unconscious processes dominate cognition. This is basic in cognitive science. Consider a simple task that seems completely conscious, like counting the number of words in a line of text. There are certainly conscious elements. You see the words and count them. But seeing the words and counting them both involve a lot of unconscious processing. Seeing the words means interpreting squiggles on a page as letters and putting those letters together to identify words. Then you have to decide what the beginning of the line is; you have to choose a starting point. Were you aware of how you chose the first word to count? And counting requires that you have an ordering of numerals in your head, can map them one to one with the words that you have identified, and then report the largest numeral when you are done. This involves a fair amount of processing. Most of it you were not aware of. Your brain did it for you.

The great psychoanalyst Sigmund Freud was right when he appealed to the huge reservoir of unconscious processes that guide human behavior behind the scenes. (There is almost no evidence, however, that he got the key processes right. Only serious Freudian devotees still subscribe to notions like boys becoming infatuated with their mothers and seeing their fathers as rivals [the Oedipus complex].) The unconscious is a powerful computer that helps us to identify people and objects, see patterns, respond reflexively, and have intuitions. We are not aware of most of the information processing we do.

Our mental computer is not only powerful, it is fast. A study reported in the journal *Judgment and Decision Making* explored how quickly people make decisions about which snacks they prefer.[1] The study had three phases. In phase 1, 12 hungry students were asked to rate how much they liked 50 different familiar snack food items like Snickers and Doritos. In phase 2, the students were shown a long list of pairs of snack foods (750 pairs to be exact). Their task was simply to look at the one item of each pair that they wanted. The experimenters tracked their eye movements so they could see where they were looking. Finally, in phase 3, one pair of snack foods was randomly selected and the student could eat whichever snack they had chosen. Subjects knew that phase 3 was coming. Its purpose was to induce them to do their best in phase 2 and look at whichever item they actually preferred. Most of the time, the students looked at the snack that they had rated higher in phase 1. Furthermore, they were more likely to look at the higher-rated snack if the difference in ratings between the two snacks was higher. So subjects were indicating real choices by looking. After subtracting the time required to register the products visually and initiate an eye movement, the researchers found that subjects made their choices in just over three hundred milliseconds. That is less than one-third of a second.

In sum, people can choose between two snack foods with incredible speed—about as long as it takes for cognition to perform one cycle of processing.[2] This result suggests that whatever cognitive processing went on in this study must have been unconscious. Conscious processes require multiple cycles of processing and so would not be that fast. It appears that if we come up with a reason, the reason is generated after the decision, not before. Such a process does not appear highly consequentialist. That is, people are not taking the time to imagine the consequences of each option; they are operating too fast to be doing that. They might have learned the consequences previously, though, and we consider that possibility below.

Unconscious Processing Can Lead to Better Decisions

The judgments and decisions we make unconsciously are often better than the ones we make consciously. Social psychologist Timothy Wilson spent much of his career at the University of Virginia demonstrating this. In an early study, people decided which of five posters they wanted.[3] The posters included a couple nice impressionist paintings and a few cartoons (anthropomorphized animals of the cool cats with sunglasses variety). There were two groups of subjects. One group essentially chose a poster and left. When they were contacted three weeks later, they were relatively happy with their choice and enjoying the poster. The other group was asked to think about their choice for a few seconds before choosing. People in this group were less satisfied three weeks later.

In another set of experiments, students were asked to either deliberate or not deliberate about other kinds of choices.[4] In one case, they chose among jars of jams. In another study, they chose among courses to take in the upcoming year. In both instances, those subjects who did not think made choices that paralleled the recommendations of experts. Those who thought consciously about their choices by deliberating made more idiosyncratic decisions that did not line up with the experts. If you take expert recommendations seriously, deliberation produced poorer decisions.

There are other reasons to believe that thinking isn't good for you. In one study, people who tended to ruminate were more likely to suffer greater depression and anxiety over time, while those who engaged more with distracting thoughts became less depressed and anxious.[5] Deliberating about your problems just makes you unhappier. Thinking about something else improves your mood.

Thinking consciously is closely related to talking aloud. It turns out that talking aloud interferes with a variety of cognitive tasks—a phenomenon called verbal overshadowing. Verbally describing an event makes it harder to remember precisely what occurred, suggesting that eyewitness informants should try to watch and not talk when encoding an event.[6] Verbal overshadowing also makes it harder to solve insight problems. Insight problems are aha problems—the kind of problem that you can think about for a while without getting anywhere until the answer pops to mind (e.g., How many animals of each kind did Moses load onto the ark? Hint: The answer is not two). It turns out that asking people to verbalize the process of finding solutions to insight problems leads to fewer solutions.[7] Furthermore, people find it harder to recognize faces and voices if they have verbally encoded them.[8]

Why does conscious thought and speech get in the way of thought and memory? The answer is that the capacity of conscious thought is less than that of unconscious thought, and this causes us to ignore important aspects when thinking consciously. We can only think consciously about a small number of things at a time. There is a meme in cognitive psychology stemming from a classic paper from 1956 by George Miller. It states that the capacity of short-term memory is seven plus or minus two.[9] One theme of Miller's paper is that we can only hold a small number of items in short-term memory, although Miller's real point is that the actual number is flexible because we can hold much more if we group items meaningfully (this is called chunking). Later work agrees with Miller that we can only think about a small number of distinct, unrelated things at any given time, though the number is likely appreciably less than seven. How many things we can think about depends on the modality we are using to think (e.g., vision versus audition) and the amount of time we have to remember them for.[10] But the point is that the capacity is limited. To think about anything consciously means to focus on some aspects of it and ignore others.

Because we have to choose which aspects to think about, we will inflate the importance of some attributes at the expense of others. Which attributes will we focus on? For one, we will concentrate on the attributes that we think are most relevant to our goal. Yet we don't always get this right. If we are trying to encode a face, we might focus on its distinctive attributes (like a particularly prominent nose) and miss the attributes that really do help us remember—for instance, holistic properties of the face like how the distance between the eyes compares to the distance from eye to mouth. If we are trying to solve an insight problem about arks, we will encode attributes related to arks in the Bible, without sufficient attention to the names of biblical characters.

Conscious thought also often focuses on justification-serving attributes. I discussed these in chapter 8, when I noted that people like to generate reasons that serve as justifications for themselves and others. In consumer choice, price frequently serves that role (I bought that one because it was the cheapest). But we can't always rely on inexpensiveness. When buying, say, an engagement ring, you would do better by focusing on an attribute that gives the item meaning.

Generally, we justify ourselves with language, so attributes that are easiest to express in language are the likeliest to be selected, all else being equal. In Wilson's poster experiment, subjects who thought before choosing chose the funny animal posters whereas those who did not chose the impressionist paintings. The beauty of Claude Monet and Vincent van Gogh is not easy to articulate; it requires a degree in fine art to acquire the appropriate language. Jokes can be easily justified because everyone enjoys a little humor, though, so they become more appealing when one has to deliberate. Unfortunately, what appeals to deliberation is not what stands the aesthetic test of time.

A succinct description of conscious thought has been attributed to German poet Johann Wolfgang von Goethe: "He who deliberates lengthily will not always choose the best."[11] The point is not to avoid being mindful when making decisions. Mindfulness is a different application of conscious thought. It does not refer to the process or strategy that one adopts in order to make a decision. It refers to awareness at a more metalevel, to being aware of the fact that one is making a decision along with the thoughts and feelings associated with making that decision, whatever they are. Those thoughts and feelings may not be causes of the decision as they are in conscious decision making; they may be effects of the decision. There is a lot to be said in favor of being mindful at all times, whether one is making a decision or smelling a rose. Being mindful can help us achieve some comfort with the decisions we make, even if we have made them for reasons we are not fully aware of.

The Role of Emotions in Decision Making

In this book, I have presented sacred value and consequentialist frames as tools that decision makers use to construct a strategy to make a choice. If that were their sole purpose, they would only be useful for conscious decision making. Are the frames irrelevant for understanding unconscious decision processes? To answer this question, we have to delve deeper into how unconscious processes work.

In chapter 5, I argued that LLMs (large language models, the stars of current AI) have some key human properties. In particular, they operate by inferring sophisticated correlational knowledge from a huge database of information that they have encoded. They pick out what goes with what in the events that they are presented with and use what they learn to answer questions by generalizing to new situations. They are trained to predict the next item in a sequence and excel at figuring out how to use what they have seen to predict successfully. Essentially, they are formidable pattern recognizers. They are trained on a vast amount of data, a big chunk of the entire internet. Obviously, humans have far more limited exposure. But people seem to operate according to the same principles as LLMs, at least at the unconscious level. We also excel at picking out what goes with what and completing patterns.

Humans do have a capability that is not found in LLMs. We are capable of conscious thought or what I have called deliberation in this book. Despite some newsworthy claims to the contrary, like when a Google engineer claimed their AI was conscious (the engineer was later fired), there is no evidence that LLMs are conscious and every reason to think they are not.[12] For instance, not one LLM has ever initiated a request for more electricity or spontaneously complained that it is bored. More to the point, LLMs confabulate outrageously by making stuff up (to verify this, try asking one to write your biography). They do not have a way of correcting those errors without further prompting. People do. People can deliberate using conscious thought to try to verify a claim before making it, and even suppress it if it cannot be verified. This is not to say that people always verify before they speak. Indeed, it is the exceptional person who is so careful.[13] But we do have the capacity to be reflective; we just do not always deploy it. LLMs do not have the capacity. It may be possible to give them the capacity by, for example, hooking up two LLMs together and having one function to verify the claims of the other. Currently, however, individual LLMs themselves are not capable of the kind of conscious, reflective thought that people are capable of and that is needed to verify claims.

LLMs also differ from people in that they do not have bodies, at least human bodies. This matters because bodies do more than consume cups of coffee and sit in armchairs. They also are intimately involved in unconscious decision making through the experience of emotion. LLMs cannot experience the same emotions as people because they do not have the necessary physical hardware to do so. At best, LLMs only capture the part of unconscious decision making that involves cold calculation, not the part that involves hot emotional responses.

Before proceeding, a disclaimer: There is not a clean distinction to be made between cold cognition and hot emotion. Any theory of emotion worth its salt has to account for the fact that emotional experiences can involve thought. For instance, when we are angry, we are angry about something, and generally the angry person knows what they are angry about. When we are fearful, we fear something (a snake, a rejection, a bad joke). We have to mentally represent those objects of our emotions and that is a cognitive act. This has been known since psychologists like William James started theorizing about emotions in the nineteenth century.[14]

But when making a decision, there is often a response that is clearly beyond the cognitive. Nerves might jingle, heart rates increase, and parts of the brain might turn red. I will refer to these bodily reactions as *affect*. The mind can be aware of them, but they are carried in the nervous system—by the brain, spinal cord, and other networks of nerves—to the entire body. That is why sweaty palms, also known as a galvanic skin response, are indicative of an emotional response.[15] And that is why feelings of anxiety can show up in the neck, back, stomach, or elsewhere. The entire body is involved in processing information.

Emotions are involved in decision making in many ways. One important way we have already seen in some detail throughout the book. Sacred values can be triggered by emotional responses, especially outrage and anger. But there are others. Jennifer Lerner, a well-known social psychologist at the Harvard Kennedy School, and her colleagues authored a wide-ranging, engaging review of the role of emotions in decision making.[16] One class of emotions that they identify are what they call integral emotions. These are emotions that arise from the judgment or choice at hand. Examples of such emotions include sympathy that causes you to donate money and a fear of falling that causes you to stick to the bunny slope when skiing. In both cases, an emotional response (sympathy or fear) is governing what you choose to do.

Integral emotions are known to be critical for sound decision making. Patients with damage to their ventromedial prefrontal cortices (vmPFC) show both flat affect and poor decision making. In the book *Descartes' Error*, Antonio Damasio discussed such patients, including the classic case of Phineas Gage, who, as foreperson of a group laying down railroad track in Vermont in 1848, was unfortunate enough to have a 3-foot-7-inch-long, 1.25-inch-diameter tamping iron blown through his skull by an accidental dynamite explosion.[17] After exiting his skull, the tamping iron landed 100 feet away. Legend has it that he stood up immediately after the accident and answered questions about what happened. It was only later that he was diagnosed as "fitful, irreverent" and not displaying normal emotional responses. Reconstructions of his skull show that the primary area of his brain damaged was the vmPFC.

In more recent studies done in the laboratory, vmPFC patients have been shown to repeatedly select a riskier financial option over a safer one, even to the point of bankruptcy. And they do so despite understanding what is happening.[18] Damasio and his colleague Antoine Bechara attribute this to the somatic marker hypothesis. The idea is that in normally functioning people, the soma (the body) experiences an emotional response to a stimulus before any thought processes occur. This response is learned via emotional conditioning that occurs (nonconsciously) due to signals from the brain, specifically the vmPFC and amygdala. The response can be described as visceral and thus delivering (literally) a gut feeling about the situation. Most rapid decisions are made based on this gut feeling without any further thought required. Further thought is possible, however. According to the somatic marker hypothesis, deliberation or conscious thought occurs elsewhere, in the lateral prefrontal cortex and other cortical regions.

The critical claim to explain the behavior of vmPFC patients is that their emotional conditioning system is damaged. This leads to a disconnect between knowledge and behavior. They do not have the feelings necessary to guide action. In particular, they lack the somatic markers that function as an automated alarm signal (e.g., a typical person's response to seeing a large vehicle headed straight for them) by forcing attention to negative outcomes. They also lack the positive somatic markers that constitute incentives (e.g., my response to the smell of cookies baking). In normally functioning people, somatic markers allow certain courses of action to be rejected and others to be pounced on immediately, reducing (sometimes drastically) the number of alternatives that need to be considered. VmPFC patients lack this decision-making pathway. All they have is conscious decision strategies, and it renders them dysfunctional. Damasio's conclusion is that the absence of emotion can be damaging to rational decision making.

Back to Sacred Values and Consequentialism

The somatic marker hypothesis states that the vmPFC is responsible for emotional conditioning. More recent work has provided evidence that the vmPFC is involved in cost-benefit calculations that are the heart of consequentialist decision making.[19] Whatever the role or roles of the vmPFC turn out to be, both of these interpretations suggest that patients with vmPFC damage are less likely to engage in consequentialist decision making. This increases the likelihood that they are framing their decisions in terms of sacred values or that sacred values serve as the ultimate justification for what they decide. If this is right, then we are forced to conclude that people who have their emotional conditioning systems impaired rely less on consequentialist reasoning. This suggests that some emotional reactions—like fear—may trigger consequentialist deciding.

We already know that some integral emotions—like anger and disgust—also influence decisions, likely through an association with a sacred value. Sympathy is associated with the sacred value "help others in need," and so we act accordingly. The association could be concrete or implied. If it is concrete, then the emotion of sympathy causes you to activate the sacred value so that it is directly mediating your willingness to give. You bring to mind the sacred value either consciously or unconsciously, and that is the reason you might donate to a charity. If the association is implied, then the relation between sympathy and donating is so strong that the sympathy itself is sufficient to cause you to donate. In that case, the sacred value may not be activated at all. Nevertheless, the sacred value can be critically important because it justifies the link between sympathy and donating. If you did not have the sacred value, you would never have built the association between sympathy and donating. For instance, the association of sympathy to the sacred value might have been concrete so many times in the past, with you donating as a result, that the link between sympathy and donating has become direct. The sacred value was critical for learning the association but is no longer necessary.

One study looking at the brain regions involved in sacred value decision making tested supporters of the radical Islamist group Lashkar-et-Taiba (Army of the Righteous).[20] Their brain activity was measured while they indicated their willingness to fight and die for sacred (e.g., "Prophet Muḥammad must never be caricatured") as well as nonsacred values. Not surprisingly, the participants expressed a greater willingness to fight and die for sacred than nonsacred values. They also responded more quickly to sacred than nonsacred values. Furthermore, the sacred value condition showed less activation in neural areas previously associated with consequentialist reasoning. In contrast to other studies, however, this one found some evidence suggesting that sacred values processing occurred in the vmPFC. This confuses the story as it is inconsistent with the idea that the vmPFC is exclusively involved in consequentialist reasoning. I will have to leave resolving this issue to others.

Emotions play other roles in decision making as well. There are incidental emotions—those that a decision maker just happens to be experiencing. These too can influence choice. Someone who enters a situation angry might decide to blame innocent individuals (the kicking-the-dog effect).[21] People in a good mood tend to be optimistic and those in a bad mood more pessimistic.[22] To illustrate, people judge the number of fatalities caused by, say, heart disease to be higher if they are in a bad mood.[23] Even the weather outside can influence how satisfied people are with their lives overall (life is better when the sun is shining).[24] Economists have even shown that the amount of sunshine on a given day predicts stock market performance in twenty-six countries.[25]

In all of these cases, incidental emotions that are unrelated to the decision at hand affect judges and decision makers. They could affect people in different ways. They could alter a consequentialist decision procedure by bringing to mind either more costs or benefits depending on whether the emotion is negative or positive. Alternatively, they could influence the likelihood of a sacred value coming to mind. Perhaps a bad mood causes people to think of sacred prohibitions whereas a good mood causes people to think about sacred values related to freedom. The effect of incidental emotions could be carried by either decision frame.

Reinforcement Learning

The somatic marker hypothesis states that emotional conditioning is a prerequisite for high-quality decision making. Because we have to make so many decisions every day, and because some of those decisions have to be made quickly, we cannot rely on slow, conscious thinking processes. Instead we rely on our bodies. Our bodies use pleasing emotions to guide us toward actions and disagreeable emotions to guide us away from actions. Our bodies learn to do this through a process called reinforcement. That process also provides another route through which sacred values reasoning can sneak into our decision making.

Reinforcement learning (RL) is the modern, souped-up version of what used to be called conditioning in the behaviorist era of psychology in the mid-twentieth century. Behaviorists tried to explain all behavior in terms of learned associations between stimuli and behavioral responses. In the case of classical conditioning, the behavioral responses were natural, built-in responses like a dog's tendency to salivate in the presence of food. In one famous case, Russian psychologist Ivan Pavlov paired a bell with food so that dogs learned to salivate in response to the bell. In Pavlov's study, food was what's called a primary reinforcer. The bell was a novel stimulus that became predictive of the primary reinforcer.

Another kind of conditioning is operant or instrumental. Here, an action is made more or less likely by pairing it with a positive or negative outcome. For example, a rat might be trained to push a lever to receive food or refrain from stepping on a platform because if it does, it will receive an electric shock. This is pure consequentialist decision making.

RL has been the object of vast amounts of study in recent years. Two systems for RL have been developed. One involves learning a model of the environment. For instance, if you were learning to navigate the streets of Moscow, you might construct a map that would allow you to navigate between places you had never been. The map is a model of Moscow in that it captures one aspect of the city, its geographic layout. Building a model while learning makes flexible decision making possible because it gives the learner the ability to respond, not just to questions that have been asked previously, but to any question that the model can answer.

The other kind of RL learning is model free. It involves learning which responses are effective given the current state of the world without any model as a guide. You might learn, say, to make your way through Moscow by knowing where to turn at each point on your route without ever having a big picture—a model—of your location relative to the rest of the city. It turns out that different neural systems support each kind of RL.[26]

RL systems are designed to select the options and actions that deliver positive reward. To learn, they make use of the neurotransmitter dopamine. Neurotransmitters are chemical messengers in the brain. Dopamine communicates what's called reward prediction error.[27] When an organism acts, dopamine cells in the brain (neurons) relay information about whether the act delivered rewards. The neurons, however, do not tell the brain how much reward was received. Instead, they tell the brain whether the amount of reward it predicted was correct. The reward prediction error is positive if the brain predicted less of a positive outcome than it received, zero if its prediction was dead-on, and negative if it predicted more reward than the outcome actually delivered. The brain then uses this information to adjust its connections so that it will make a better prediction next time. The brain appears to be a machine designed to learn by making predictions and finding out if they were correct or not. We saw in chapter 6 how critical the ability to predict the effect of actions is to good decision making.

This is also how many modern machine learning models are trained. They are trained to predict the next word in a text, and their knowledge base is modified until their predictions are successful. In some cases, they are trained using the same mathematical rules that are used to model RL in the brain. The learning process in all machine learning applications is consequentialist. Learning is understood as the application of rules of choice based on knowledge about what choice options deliver—knowledge acquired through previous learning experiences.

Decision Making Without Control: Habits and Addictions

The RL mechanisms that I have briefly reviewed offer an account of how we acquire the most common strategy for decision making, habit. Most options are chosen simply because that is what we have always done. We put our right shoe on first because it has proven effective in the past. In the language of RL, we have put our right shoe on first in the past and been rewarded, so we have built a stronger and stronger association between the desire to have shoes on our feet and the action of putting the right shoe on first. This habit can become sufficiently ingrained to become a superstition. You might come to believe that if you do not put your right shoe on first, bad consequences might befall you.

There is good reason to believe that the mechanisms that cause addiction operate in exactly the same way as those that learn habits. In the case of substance abuse, we learn that the cigarette, drink, or injection delivers a reward, and so we learn to consume the substance. A parallel process occurs for gambling or screen addiction or any of the activities we become addicted to. What addiction teaches us is just how powerful RL-learned associations are. They are so powerful that they cannot be overcome even by the knowledge that they are doing us tremendous harm and a strong conscious desire to make a different decision. The power of these associations is evident in other kinds of activities too, like workaholism and eating disorders. They even manifest in habits that we might be proud of. Some of us cannot prevent ourselves from exercising or doing what is necessary to get good grades. The only difference between these desires and what we normally call addiction is that our conscious minds are not opposed to these habits.

There have been multiple theories of addiction over the years. The disease model holds that addiction is an illness that comes about as a result of the impairment of healthy neurochemical or behavioral processes. RL offers one version of this view. It states that disruptions in dopamine circuits can lead to impulsive behaviors.

Another theory is the opponent-process model. It assumes that every psychological event A will be followed by its opposite event B. For instance, the pleasure one experiences from heroin is followed by an opponent process of withdrawal. And to avoid withdrawal, people desire more heroin. The reason for opponent processes is that the body has a natural tendency to maintain balance or homeostasis. So a chronic elevation in levels of dopamine will result in a decrease in the number of dopamine receptors available. This is called downregulation. Because fewer dopamine receptors are available, neurons become less excitable, leading to weaker reward pathways. The fact that reward pathways are dulled would explain addicts' difficulty feeling pleasure (anhedonia), their increased tolerance for the substance, and withdrawal symptoms.

A final theory I will mention is behavioral economist George Loewenstein's visceral theory of addiction.[28] It is like the opponent-process model, although it alleges that addiction does not result from the desire to avoid withdrawal but instead from craving. Addiction, on this view, is like appetite. When we are hungry, we need to eat, and when we are thirsty, we need to drink. In other words, addiction depends on somatic markers. It results from visceral factors playing their normal role, serving as interrupts to make us focus on high-priority goals.

The visceral theory assumes that cravings are conditioned to one's environment. If one usually has a cigarette after dinner, then one becomes conditioned to craving a cigarette when one is finished eating, presumably through an RL process. This differs from the opponent-process theory, which claims that it is withdrawal, and not craving, that is conditioned. As anyone who has experienced intense hunger or thirst can attest, the price of craving is that it focuses us on the present. We cannot think about anything except satisfying the craving immediately. So today's desire crowds out tomorrow's, and we become willing to trade off satisfying the craving for costs down the road. In extreme cases, the costs eventually paid include all of the addicts' assets and even their lives. The other cost of craving is that it causes us to focus on ourselves and neglect others. Torturers, for example, understand that depriving someone of sleep or nourishment leads to cravings so powerful, they can cause betrayal of friends and family.

Return to the Community

Unconscious decision processes are usually understood as occurring within the individual decision maker. Emotions occur within the individual and the RL processes we have discussed occur in the brain. But that perspective ignores the critical and constant influence that other people have on us, and the fact that others often make decisions with and for us. While reinforcement can come from success or failure at a task that has nothing to do with others, we are frequently reinforced by other people. Others might pat us on the back, smile, or give us a raise for making the right decision. Alternatively, they might shake their heads, avert their eyes, or fire us for making a decision they do not like. We might even reward ourselves for knowing that others would approve or punish ourselves if we believe they would disapprove even when others don't know, or don't know yet, what we have decided. That is why we might feel pride or guilt for a decision we make that affects others even when we make the decision anonymously.

One source of reinforcement is the reaction of our community to the sacred values that we express and act on. Red state governors are rewarded for saying, "Family is first," and blue state governors are rewarded for saying, "We support diversity and inclusion." Professional athletes get brownie points for saying, "There's no 'I' in team," and professors are loved for putting compassion ahead of discipline. Some people elicit positive feedback for expressing their sacred values on a regular basis. Such people are called honorable and principled by those who want to reinforce them, and sanctimonious, self-righteous, or virtue signaling by those who either disagree with the values or doubt the sincerity of the display.

A critical function of our complex emotional lives is to mediate this complex of social reinforcement. One of Jennifer Lerner and colleagues' contributions was to identify the several roles that emotions play in social decision making. One role is to help people understand one another's emotions, beliefs, and intentions. My ability to feel anger allows me to empathize with the anger you feel if, say, someone insults you. That gives me some purchase to know what you are thinking about that person, and helps me understand your intentions if you lash out. People are quite sensitive to others' emotional expressions.[29] This allows us to recognize and interpret reinforcement signals we get from others. A second role is that feelings like outrage cause us to punish people for misbehavior. And the knowledge that others might punish an action influences which actions people take. Third, emotions in us can evoke complementary, reciprocal, or shared emotions in others. For instance, communicating gratitude or anger can trigger others to do the same.[30] So not only do emotions support decisions that we make personally and privately but they are also intrinsic parts of the decisions we make with and concerning others. They help us decide what feedback is appropriate and help us make sense of the feedback we receive and the feedback we observe others receiving.

So social reinforcement comes in a complex pattern that explains both how we internalize consequences and sacred values. As far as the brain is concerned, they are instilled through individual and cultural experience via the same mechanisms that instill habits and addictions. Sacred value and consequentialist frames provide different decision-making strategies. We have seen that those different strategies have different consequences for society. But people acquire both through the same neural mechanisms.

11 How to Do the Right Thing

Cameron Todd Willingham's house burned down in Texas in December 1991, killing his three young daughters. A police investigation determined that the fire had been intentionally set. Willingham himself quickly became a suspect and eventually was accused. He was tried and convicted of arson and murder in 1992.[1] Willingham insisted that the fire was accidental, and maintained his innocence throughout the trial and afterward. He even turned down a deal of a life prison term in exchange for a guilty plea that would have taken execution off the table. Partly on the strength of claims by medical experts that Willingham's tattoo of a skull and serpent fit the profile of a sociopath along with a statement by a psychologist that his poster of the heavy metal band Iron Maiden and another of a fallen angel from the rock band Led Zeppelin was an indicator of "cultive-type" activities, Willingham was sentenced to death and executed by lethal injection in 2004.

There is good reason to believe that Willingham was in fact innocent. The evidence against him was effectively rebutted in 2004 by fire investigator Gerald Hurst. In 2009, a report by the Texas Forensic Science Commission found that the arson investigation in Willingham's case was deeply flawed.

If Willingham was indeed innocent, then his decision to refuse a plea deal was likely driven by a sacred value. He said as much. Willingham insisted that he would not admit to something he had not done. Apparently, doing so would be tantamount to violating a fundamental principle of his. This principle goes beyond not telling a lie. The lie in this case would be a public statement that he had done something he had not. But in this case it would require a public admission that he was a monster. He felt so strongly about it that he risked his life and, in the end, lost it. Dilemmas like this are not uncommon when accused but innocent perpetrators are offered plea deals. Should they violate a sacred value or risk a serious consequence like death or prison?

Although most of us are not faced with such stark and weighty choices, we do sometimes have to decide whether to violate a sacred value to avoid a negative consequence. Thus we tell a white lie to avoid hurting someone's feelings or skip going to religious services to watch an important football game. What should be our guiding star? Should we have faith in sacred values and let them govern our decision making? Or should we be consequentialists and make decisions by doing cost-benefit analyses on outcomes? Let's start by reviewing the pros and cons of both.

In Support of Sacred Values

In chapter 2, I argued that sacred values are essential. Few of us would befriend someone who had no sacred values. A person's sacred values are fundamental to their identity because they represent central aspects of how they interact with the world—which actions they consider most justified and which actions they consider anathema. Moreover, a person's sacred values signal what community they associate with. Communities that hold social and political views are brought together and maintained by their sacred values. You cannot be a genuine member of the Proud Boys if you do not support white supremacy, and you cannot call yourself a left-wing progressive if you do.

Sacred values certainly change over time. The connotation of "all men are created equal" in the Declaration of Independence has morphed in meaning in the United States from "all landed white men are created equal" to "all landed white people are created equal" to "all people are created equal." At any given point in time, there is a core, prototypical sense of the phrase that is a part—usually a central one—of the identity of those who subscribe to it. For most of us Americans today, an indispensable part of our identity is that we accept the interpretation of the phrase that dominates at the moment. For most of us, a requirement of anyone we would refer to as a friend is that they do so too.

Many sacred values shape our identity as members of a community. Citizens in my community must bow to a host of dictates: They must value not harming other people either physically or emotionally, believe in not harming the planet, and value facts over obvious untruths. Anyone who rejects one of these sacred values would not be welcome in my social circles.

Then there are other sacred values that lie at the edges. Many in my community would treat someone as a pariah if they don't value scientific evidence over hearsay. Once a finding has gone through the rigorous, adversarial process of peer review, if it is published in a credible journal, then it is as definitive as any other source of information short of direct experience.

This is a more problematic sacred value in part because not all scientific evidence is the same. Any social scientist knows that most measurements of people and their societies are chock-full of uncertainty and many findings are hard to replicate. Part of the problem is that some of the work that we call science is actually done to allow a student to graduate or for the sake of a promotion and not solely in an objective pursuit of truth. Many "scientific" findings need to be taken with a grain of salt.

Sacred values like reverence for scientific evidence have limited application and could require a fair amount of analysis to decide if they bear on a particular case. This is directly analogous to the law. Legal scholars devote a lot of time and effort in some cases to determine which legal rules apply. The question of whether patients have a right to die became national news in 1998, when the husband of Terry Schiavo petitioned in court to remove Schiavo's feeding tube. She was in a persistent vegetative state due to brain damage caused by cardiac arrest. Schiavo's husband claimed that she had told him that she would never want to be kept alive artificially, but her parents fought the removal of the feeding tube in court. Eventually, the judge decided in favor of the husband, and Schiavo's feeding tube was removed and she died in 2005. The issues in this case included who has a right to die as well as questions of fact, like whether Schiavo had any potential for recovery and whether her husband's report of her preference to die was credible. The question of how to decide if a law applies to a particular case can be involved and can engender disagreement. The same is true of sacred values.

Theologians can also devote years to deciding which sacred values apply to a particular situation. Christians have debated for centuries what is truly holy, whether the host is the actual body of God or just a sign, and whether God is singular or a holy trinity. As a result, communities differ in which sacred values they accept, and there are differences even within communities. There are those in my wider community who favor religious principles over science. I will reveal that these are not my closest friends, but I would happily invite them for a drink or discussion. Some are family, and I engage with them frequently.

Communities may be governed by sacred values, but the landscape of communities is complex. For instance, different denominations of Christianity are nested within the larger community of Christians. The broader community accepts a broad sacred value, but it gets specified in different ways by the nested communities. In some cases, communities overlap with other communities. Those who are leading the fight against anthropogenic climate change, for example, can be found across communities of religious as well as nonreligious folk, and within every political party. Sacred values can cut across tribal perspectives.

Against Sacred Values

Despite all of these virtues, we know that all is not hunky-dory with respect to sacred values. Indeed in chapter 8, I reviewed evidence that we usually oversimplify, and that one way is by relying on sacred values. They cause us to jump to conclusions that channel our communities rather than engaging in serious critical thought.

We have seen other costs of sacred values too. Sacred values are defined as absolute, not amenable to trade-offs, and thus they come with a certain rigidity. Rigidity can be admirable, as it is when people die for their values, especially when we share those values. But most of us do not admire those who prevent their children from getting decent medical care because blood transfusions or other forms of medicine violate their sacred values. Most of us also consider it overly rigid when people refuse to protect the environment, claiming that it is God's dominion.

Part of the problem with sacred value frames is that the unwillingness to engage in taboo trade-offs bleeds into tragic trade-offs. Recall that a taboo trade-off prevents one from violating a sacred value in favor of material gain. One should not murder no matter how much money you will get for doing it even if there is no chance of being caught. A tragic trade-off is a trade-off among sacred values. One might have to kill for the sake of defending one's own life or the life of another innocent. One might have to skip prayers in order to help out in an emergency. Sacred values can come into conflict, and it might be impossible to avoid violating all of them.

Tragic trade-offs can be forced on us. The real tragedy is that we sometimes mistake tragic trade-offs for taboo ones. This strikes me as an error made by a small number of people opposed to abortion. It is certainly reasonable to believe that abortion is murder and easy to respect the position that refusing to murder is a sacred value. When a person's life is at stake unless they have an abortion, however, then prohibiting abortion itself becomes a weapon that could kill. In this case, for a believer, there is a tragic trade-off. Either the fetus or mother has some chance of dying. Laws that prohibit abortion under any circumstance, including cases where the mother's life is at stake, fail to recognize this tragic trade-off, thereby preventing a fair and just process to resolve it. We may disagree on how to resolve it, but we should be able to agree that there should be some means. More generally, every sacred value on occasion comes into conflict with some other sacred value. When that happens, something has to give. Sacred values, even those that define a community, have to be subject to trade-offs. An unwillingness to consider that possibility is a form of excessive rigidity.

Problems become catastrophic when the rigidity becomes emblematic of a social movement. When communities develop around a sacred value, the rigidity becomes self-reinforcing. When different communities develop around competing sacred values, then we get polarization and its attendant dysfunction. Much of the problem is that groups become increasingly adamant about the righteousness of their own value in order to contrast themselves with the competition. Small differences get writ large because of the social dynamics of difference.

We can see this pattern in the evolution of attitudes toward COVID vaccines in the United States in 2020. When the pandemic started to break out in the United States in March of that year, Republicans' attitudes toward getting the COVID-19 vaccine were slightly more negative than Democratic attitudes.[2] This is not too surprising, as Republican attitudes toward the measles-mumps-rubella vaccine were also more negative than Democratic ones.[3] There are antivaxxers on both sides, but Democrats are, on average, more supportive of any kind of vaccine than Republicans.

But then President Trump decided to downplay the risk of the pandemic. On January 22, 2020, Trump told a reporter that the country had the coronavirus "completely under control." On February 27, while US health officials were warning that the pandemic might be long-lasting, he said a "miracle" might make the pandemic "disappear." On March 11, he said that for "the vast majority of Americans, the risk is very, very low."[4] He continued to deemphasize it through 2020.

During this period, the divergence in views about the vaccine between Republicans and Democrats grew. While Democratic attitudes stayed roughly the same from April 2020 to August 2020, Republican attitudes became increasingly negative.[5] I suspect that the reason for this decline among Republicans is that they followed their leaders. Republican leadership decided it was better for the country, or at least for Republicans, to talk and act as if the risk of the pandemic was overblown. Over the six months from March to August, this view became central to the Republican administration and its adherents. It became a sacred value. It is possible that Trump's statements and Republican attitudes are causally unrelated and there is some other reason that Republican attitudes declined. But I have no idea what that other reason might be.

In a study I did with Mugur Geana and Nat Rabb, we asked a few hundred people in September 2020 about their attitudes toward COVID-19 mitigation procedures like physical distancing and wearing a mask.[6] We also obtained a bunch of demographic and personal measures (age, sex, race, knowledge about coronavirus transmission, whether they knew anyone with COVID, whether they worked with patients, etc.). We also asked our participants about their political leanings. The best predictor of their attitudes toward physical distancing and mask wearing was participants' political leanings. Overall, knowing whether they were Republican or Democrat was more informative about their attitudes toward COVID than whether they had knowledge about how COVID transmits, their personal experience with the disease, or anything else. In another couple of surveys that Mae Fullerton administered in April and October 2020, we found that ideology was a better predictor of COVID mitigation actions even than whether an individual had risk factors that made them susceptible to serious respiratory disease.[7]

The early period of the coronavirus pandemic was an exceptionally uncertain time. Nobody knew for sure how deadly the disease was, and even experts did not know exactly how it was transmitted. Many of us spent a lot of time washing our hands because the authorities (the Centers for Disease Control and World Health Organization) told us to. Although we may have benefited from having clean hands, it probably did not protect us from COVID.[8] It took a couple of months to become clear that wearing protective masks would be beneficial. Fear ran rampant. There was certainly a lot of talk of causes and consequences, but most of us did what people do when they are highly uncertain and scared. We turned to other people and took our cues from their behavior. And it turned out that when we turned, we saw one of two Americas. Some of us were surrounded by Democrats who took the scientists' warnings and recommendations seriously, so that is who we saw. And some of us were surrounded by Republicans who had doubts about the credibility of elitist scientists and thought pandemic fears were largely due to fearmongering.

To the extent your own views and actions were determined by your community, you were not engaged in consequentialist analysis. You were being controlled by your communities' sacred values—the rules of action that it agreed on that had become symbols of its identity.

The science of disease mitigation was certainly about consequences. To the extent a group was influenced by the science, it was being influenced by consequences. But to the degree an individual was willingly doing only what those around them did, any appeal to consequences was indirect. And the fact that ideology was the strongest predictor of attitudes and behaviors suggests that the appeal to others was not only common, it was paramount.

The fact that two communities increasingly parted ways in their view during the early pandemic suggests either that their sacred values became more and more important over time, or that each group's values became more and more differentiated from the other group's. Either way, the groups' sacred values had become increasingly emblematic, and this led to greater polarization and hostility.

This dynamic has occurred with many values. In the United States, it explains what has happened with gun control, transgender rights, and abortion. In each case, a small difference in values has grown so that each side's perspective is more and more homogeneous, differentiated from the other side's, or both. And in each case, each community's sacred value has become emblematic. Expressing your view on the issue becomes equivalent to waving your identity group's flag. This dynamic is not limited to the US. It happened in the United Kingdom in 2016 over Brexit and in Germany in the 1930s over Nazism; it is happening in Brazil over a number of issues including crime and in South Africa over corruption. There is probably not a nation on earth that has not experienced it.

Sacred values, like habits and addictions, are inculcated in us through repeated exposure. They become ingrained through a learning process that depends on obtaining a reward, which can come in the positive form of social approbation or negatively as social disapproval. The fact that sacred values are so strongly associated with emotional reactions like outrage and anger suggests that the conditioning through exposure is emotional. In this sense, those sacred values that we have held onto for a period of time can be thought of as cognitive habits. When we deploy them, we do so because of our history of conditioning within our community. They are not the outputs of slow, deliberative thought. This gives them the potential to be dangerous.

In Support of Consequentialism

I have now summarized the arguments for and against treating sacred values as our guiding star. What about the other decision strategy—consequentialism?

One way to understand the benefits of choosing based on consequences was reviewed in chapter 4. There I described expected utility theory in some detail. EU theory is the banner consequentialist theory, at the heart of most economic theory, and a serious contender in the pantheon of decision theories studied by philosophers and other decision theorists. Its prescription is to always choose the options that provide the most expected utility, where utility is a measure that reflects how much you want something. The great benefit of choosing to maximize expected utility is that it is the strategy that is most likely to deliver what you want. So if you spend your life choosing by EU theory, you are most likely to have received the most of what you want. What could possibly be a better decision strategy?

The other argument I reviewed in favor of EU theory was more abstract. I showed by example that if you do not choose according to EU theory, you can be turned into a money pump. More specifically, if you do not accept any of the assumptions that EU theory makes about preferences, but otherwise choose to maximize your utility, then a devious agent could extract all of your wealth from you. That would not be welcome, but most of us are too smart to let that happen.

Notice that EU theory is more general than it might appear. Various decision strategies fit into the expected utility umbrella. You might choose based on the attribute you care about the most. For instance, you might always choose the option that you believe is worth the most money, most likely to impress your mother, or most likely to make you comfortable. Or you might choose to minimize loss and focus only on the costs of options, ignoring their benefits. Or you might choose some hybrid of these strategies. Notice that all of these strategies deliver a utility for each option—one based on a subset of the option's attributes. As long as your strategy considers options' utilities and combines them with a measure of the uncertainty associated with getting the relevant outcomes, it is a form of EU theory.

You can still be a consequentialist even if you do not buy into EU theory hook, line, and sinker. For instance, you might just focus on the most likely outcome associated with an option. You might ignore all outcomes of your state lottery except the most likely one—that you win nothing. And you might use this as a reason not to buy a lottery ticket. Or you might do a worst-case analysis. In deciding how to invest your retirement money, say, you might choose the option that has the best worst outcome, the investment strategy that is guaranteed to leave you the largest amount even though other strategies are more likely to leave you more (but also might leave you less). These examples ignore the probabilities of outcomes; they focus exclusively on a single outcome. They are consequentialist because they choose based on outcomes, but they are not expected utility strategies because they do not calculate expected utilities. Consequentialism comes in a variety of forms. It gives you lots of ways to choose, with the only constraint being that the choice must be based on its consequences.

The other advantage of consequentialism is that it allows you to incorporate everything you know that is relevant to your choice. You can assess and evaluate consequences using all the information at your disposal. You can also use all relevant information to determine the likelihood of consequences. What will actually happen if I choose this option? You can refer to previous experience to answer that question, to others' expertise, and to your own causal knowledge. What will happen if I choose a Labrador retriever puppy rather than a goldendoodle? I can appeal to my own experience showing that Labs are smart, adoring, completely focused on food, and shed masses of fur even in the middle of winter, whereas goldendoodles are less extreme on all of these dimensions. I can ask my local dog trainer too. Or I can use my causal reasoning ability to predict that a Lab will provide more benefits (an animal that will know what I want before I do) but more costs as well (creating tension in the family as I am unlikely to ever spend enough time vacuuming up fur). Consequentialist decision strategies give you the opportunity to use your most sophisticated tools to make the best decision.

In contrast to sacred value frames, consequentialist frames allow trade-offs. In fact, they require them. Indeed, the whole notion of doing a cost-benefit analysis to make a decision is to make trade-offs. The trade-offs in consequentialism are about consequences and how to trade off the consequences associated with one option with the consequences associated with some other option. One car has great gas mileage but little storage space, and the other is better looking than Brad Pitt but tends to break down. How do I weigh the attributes within each automobile and compare one car with another? This is what multiattribute decision theory is designed to help with. Consequentialism can take everything into account in a systematic way.

Against Consequentialism

The first problem with consequentialism is that it is not always easy. In fact, applying it can be outrageously difficult. It can be difficult just because decisions themselves are difficult and the human mind is not designed to apply consequentialism optimally.

Let's consider the difficulty of decisions themselves. Decisions like whether or not to get dressed in the morning are not difficult. The benefits of getting dressed relative to the costs are so overwhelming for anyone who is not visiting a nudist colony that one can make a satisfactory decision without any work at all. And we make many decisions like that every day—thousands of them if you include whether or not to take the next breath, or which foot to put ahead of the other when walking. But even apparently simple decisions can be difficult. Should I eat that second piece of cake after dinner? Here I have a conflict between the short-term benefit of enjoying the cake versus the long-term benefit of not overeating. Which pair of socks should I wear? If all of your socks are similar, then all options are essentially equivalent. This is an example of the proverbial donkey caught between two haystacks. One way to deal with it is to leave the decision to chance. Put your hand in the drawer and see which pair you come up with.

Many decisions that matter are really difficult. That is because it is often difficult to determine what the different options entail. There is both causal complexity and uncertainty. Causal complexity is a challenge when dealing with artifacts, biological systems, or social issues. Consumer choices require predicting the functions, durability, and longevity of the different options on offer. I still have not figured out all the functions of my coffee machine and certainly cannot predict when its electronics will burn out. It just requires too much engineering knowledge and time to answer a question like this. Every consumer decision I make would take a week if I answered such questions first. And this degree of complexity is nothing relative to choices about medical preventions or treatments. Everybody has a theory about, for instance, what we should and should not eat. But try spelling out the causal pathway from the ingredients of the food to its beneficial or ill effects, and even biochemists who specialize in these issues can get stumped. As for social issues, the complexity is evidenced by the fact that the outcome of all major political events is unpredictable. Who would have guessed that the election of a Black US president in 2008 would have stirred up as much white supremacist backlash as became apparent years later?

The complexity of all of these decisions forces uncertainty. In each case, there are a host of things that the decision maker cannot know. I cannot know if my coffee machine has a microtear in some critical component. I cannot know for sure how a medical treatment that affects different people in different ways will affect me. This is called epistemic uncertainty. It is uncertainty due to limitations in my knowledge that could be overcome if I knew more. There is another kind of uncertainty, aleatory uncertainty, uncertainty inherent in the world. There are events that occur randomly and that I cannot predict no matter how much I know. I cannot know what will happen if a brand-new microbe—different from all previous microbes—evolves in a few years and finds it can survive and multiply most successfully in human hosts. How can I predict the consequences for human life if I don't know anything about the microbe? Even simple decisions like whether or not to take a breath have aleatory uncertainty. After all, in my next breath I might inhale a virus or toxin that happened to make its way to me from miles away. So uncertainty is inherent to decisions of any complexity and exists to some degree for all decisions.

The other set of challenges associated with consequentialist frames is that the nature of the human mind creates its own problems when reasoning consequentially. This was reviewed in chapters 5 through 8. First, what we want is not necessarily what we like. This is most obvious in the case of serious substance abuse. We can intensely crave a drug that no longer gives us pleasure. Second, we often do not even know what we want. Do you want success and attention or security and freedom? The fact that our desires can be pushed around by how options are described and choices are made—as we saw in chapter 7—suggests that they are labile and that properties irrelevant to outcomes affect them. Third, we cannot expect people to calculate expectations correctly because expectations depend on judgments of probability, and people's probability judgments are systematically biased. We do not have the cognitive resources to consistently calculate correct probabilities, so we often rely on heuristics. Those heuristics, like all heuristics, can lead us astray.

Another challenge of consequentialism is what to do when decision making is by a group, not an individual, and members of the group disagree about utilities or probabilities or both. Consequentialism alone does not offer a solution. Should members try to compromise? Should democracy prevail? Should expertise prevail?

Power is often the critical factor. Generally, the boss or whoever has the most resources gets the final word in the absence of a strong norm to follow a more egalitarian decision-making procedure. These are issues beyond the kind of consequentialism that I have been considering. That kind of consequentialism merely states that consequences should matter, but not to who or how.

Choosing by sacred values is no panacea for this problem. It can run into a parallel problem when different members of a group place priority on different sacred values. Indeed, I identified this as the biggest source of conflict associated with sacred values. It is a problem whenever groups are diverse enough to be composed of people with divergent sacred values. The problem is reduced by the fact that most groups are defined by their sacred values so that they share key ones. Members of a committee to save the western coyote are likely to agree that saving the western coyote should be a priority even if their sacred values differ in other ways. But such agreement is not the end of the tale. An effective committee will, at some point, turn to figuring out the best way to achieve its goal. To do that, it will have to put on its consequentialist hat. What are the costs and benefits of each approach to saving the western coyote? Some might believe that putting up fences will provide the greatest leverage, and others that raising awareness will. Sacred value decision strategies only require agreeing on broad values about action; consequentialist strategies require agreeing on detailed plans to achieve outcomes.

What to Do?

Both sacred values and consequentialism have their place. Sacred values unite us and set parameters for us to live constructively in a formidably complex world. Consequentialism gives us a way to make good decisions—the ones that have a good chance to produce the world we most want to live in.

We learned in chapter 3 that we tend to rely more on sacred values than we should. Sacred values are a way of simplifying choices and have the invidious effect of giving us a sense of understanding without true understanding. As a result, they can lead to bad decision making and play a role in creating polarized societies. People with competing sacred values too often cannot communicate with one another. Although we cannot and should not give up on sacred values entirely, we would live happier lives in less conflicted societies if we relied less on them and put more time and effort into framing issues consequentially.

But it is impractical to always be a careful consequentialist. It takes too much time and requires skills most people do not have. If we tried to compute the consequences every time we took a breath, we would suffocate. For important decisions that we make occasionally, though, we can spend more time than most of us do calculating consequences and deliberating about which outcomes we value. Doing so would lead us to rely less on habit, making our decisions far more informed and effective, and would steer us toward a society less prone to hostility and polarization.

Being Critical of Critical Reasoning Skills

How can we achieve that? The reflexive answer to this question is that we need to change our thinking styles. We need to be more reflective and engage more in critical thinking. We need to learn the cognitive tools that prevent us from deploying error-prone heuristics and making fallacious inferences.

A number of educational programs have been launched to improve critical thinking. Some use abstract problem-solving as a tool, others use mystery stories, and some focus on everyday problems. In each case, students are given examples of good critical thinking either as individuals or in a group format and then are asked to go off and practice good thinking skills on their own.[9]

A different approach attempts to improve reasoning by teaching people how to take alternative perspectives than the restricted subjective one we normally assume.[10] The idea is to encourage people to break the tendency to see and appreciate only the evidence that supports what we already believe, and to learn to attend to more than just the data that confirm our own hypotheses about the world. An early version of this technique is to train people to "consider the opposite" or "consider an alternative." To illustrate, when you are judging your confidence that the position you take on some political issue is correct, like whether or not an assault weapons ban is a good idea, you might also ask yourself why the position opposite to your own is a good idea. This will send your thought process down a new pathway and generate new ideas. It tends to reduce overconfidence to some extent.[11] A related strategy championed by University of Pennsylvania psychologist Jonathan Baron is to teach people to be actively open-minded thinkers, willing to consider alternative opinions and evidence that contradicts their current beliefs. Scales that measure actively open-minded thinking are reasonably strong predictors of performance on the kind of judgmental tasks I discussed in chapter 5, and they predict how much individuals tend to engage in superstitious thinking and conspiracy theorizing too.[12] Deploying strategies like these can be effective for reducing the cluster of biases that is frequently referred to as confirmation bias.

It cannot hurt to develop critical thinking skills. After all, in the process one necessarily learns some logic and hopefully reduces the habit of being cognitively lazy—a not uncommon phenomenon. Obviously, education itself contributes to better decision making. Being able to read, write, and do arithmetic gives one access to information, ideas, and skills that contribute to thoughtful decision making as well as reduce bias.[13] It allows one to see, for example, a fact that turns out to be counterintuitive: one chance out of ten is better than nine chances out of a hundred.[14]

But I do not believe that training critical reasoning skills is the best way to approach the problem. First, there is little evidence that it works to solve broad societal problems like polarization. Training does seem to have small beneficial effects for reducing bias on narrow tasks. For instance, there is some evidence that medical professionals can be trained to make better medical decisions.[15] These small effects can be important. Yet there is essentially no evidence that people can be trained to make better decisions across the multiple aspects of our lives. The effectiveness of programs decreases as the material being tested gets further away from the material used for teaching. This is how cognitive scientist Daniel Willingham puts it in an influential paper from 2008:

> Can critical thinking actually be taught? Decades of cognitive research point to a disappointing answer: not really. People who have sought to teach critical thinking have assumed that it is a skill, like riding a bicycle, and that, like other skills, once you learn it, you can apply it in any situation. Research from cognitive science shows that thinking is not that sort of skill. The processes of thinking are intertwined with the content of thought (that is, domain knowledge). Thus, if you remind a student to "look at an issue from multiple perspectives" often enough, he will learn that he ought to do so, but if he doesn't know much about an issue, he can't think about it from multiple perspectives.[16]

Not only are we often unable to apply critical reasoning skills in unfamiliar domains, but we usually do not apply them even when we can. Frequently what prevents good decision making is a lack of awareness that our current procedure is problematic, not an inability to do better once we try. Training often fails because people don't even realize when it is time to deploy what they have learned.[17] One reason that people fail to recognize that they have made a problematic judgment or decision is that feedback can be delayed, as it is when one supports a policy on, say, going to war in Iraq whose negative consequences only appear years later.[18]

Here is an illustration of this lack of awareness. One behavior that contributes to polarization is sharing false news stories. People overwhelmingly share stories they agree with more than those they disagree with.[19] This is true even though most people, most of the time, do not have firsthand knowledge of stories' contents (because they don't read them), and so what determines whether they agree or not is whether the headline of the story reflects their sacred values. In the United States, Republicans tend to agree with stories about how the decline of family values has led to an increase in rates of abortion and Democrats tend to agree with stories about how eliminating abortion centers endangers the health of expectant mothers. But it turns out that if people stop for a moment to consider whether or not the story is true, they are better than chance at identifying fake news.[20] A story that claims that people were, on average, getting one abortion every three months is just not likely to be true. Nobody gets pregnant that often. Anyone who devotes a little thought to the issue is likely to figure that out. So the problem is not identifying what is true and false as much as it is bothering to try. Most people do not raise their truth radar when reading news. We just assume what we are reading is true and go from there. Once we are made aware that what we are reading might be false, we are better at figuring out whether we should believe it or not. The problem is less our ability to reason critically than it is the goals that drive our thought and behavior. Instead of concentrating on winnowing out truth, we are focused on acquiring new facts, amusing ourselves, and titillating others.[21]

The final reason that I do not believe that developing critical reasoning skills is a silver bullet for society's ills is that it does not address the fundamental problem of framing issues in terms of sacred values rather than consequentially. Have I made a mistake if I frame a conversation about abortion in terms of sacred values? I don't think so. Doing so is not a case of fallacious reasoning unless I use faulty logic. I am simply framing the issue in a way that is meaningful to me. The problem with a sacred value frame is what it fails to include, not that it necessarily elicits errors. In the case of abortion, it fails to elicit thoughts about effects over time on the child, mother, and society. These are complicated and require careful, consequentialist analysis. The other problem with the sacred value frame is that it makes it hard to talk to people who do not share the sacred value. There is nothing inherently wrong with the reasoning associated with a sacred value frame, so critical thinking skills will not induce us to reframe issues nor tell us how to do it.

Solving the Problem

It is hard or even impossible to change the way lone individuals think, but surrounding people with the right cultural supports can compensate for our shortcomings.[22] What we need to do is to change social norms.

Our new social norms should dictate that those with strong opinions about a policy should have some understanding of those consequences. Most of us do not have anything resembling that for most issues. We need to be able to admit our ignorance and be less fervent in our proselytizing. In other words, we should have less hubris. We need to acknowledge the complexity of issues and the fact that our knowledge is limited, both our own knowledge and that of most others. We need norms that make us feel that simply asserting a preference—even a sacred preference—is not sufficient justification for a policy position.

Social norms that demand an understanding of how policies cause consequences can raise the temperature of a conversation. They require a willingness to call someone out and ask them to explain themselves when they are unlikely to have much of an answer. Coming across as uninformed is not pleasant and can sometimes be devastating, as it was for Libertarian presidential candidate Gary Johnson when asked a question by a journalist about a war-ravaged city in Syria in 2016; he responded, "What is Aleppo?"[23] But most of the time the experience of others showing you know less than you think is tolerable. Calling someone out gently is usually just provocative. Realizing and acknowledging that you know less than you think should not be that hard, and isn't for most people.[24] It just requires admitting you are human. The hard part is letting go of any tendency toward self-righteousness. That requires accepting that others might be correct in thinking you're wrong.

Changing social norms requires respected role models who support the norms and demonstrate them in action, broad conversations across all communications media appealing to the new norms and chewing them over, and buy-in from both thought leaders and the rest of us. There are also some individual skills that can be trained to facilitate changing norms—skills that operate by helping individuals make use of others' knowledge. First, we need to develop means of dealing constructively with conflict. If more people master techniques for dealing with conflict, then it would be easier to have difficult conversations and learn from one another. And if we knew others had those techniques available, it might be easier to ask provocative questions that test real understanding without worrying that we are going to cause strife. Second, we would all benefit from more discussion of consequences and fewer pronouncements about sacred values.

One technique for reframing an issue to focus on consequences instead of sacred values is simply to ask for a causal explanation. Imagine being told that someone you don't know well is a homophobe. Many people would react with disgust, and some perhaps with fear, and would dismiss the offender as odious or a threat. For at least some of these people, homophobia is a transgression of their sacred value that everyone should be treated with respect regardless of who they love and that violators of this value deserve to be shunned. Yet such a values-based frame can be transformed into a more mechanical, consequentialist one. Those who respond to the news with disgust might ask themselves the question (or perhaps someone else should ask them), Why is this person like this? What caused them to be a homophobe? Such a reframing does not excuse the homophobe. It simply changes the thought process and therefore the conversation from one about who is right and wrong to one about the nature of the world this person grew up in as well as how the world works.

This reframing has several effects. First, it is a natural segue into other important questions like who or what is responsible for the individual's attitudes, and what can be done about it? What is it about our society that leads to homophobia? Because we are now talking abstractly about cause and effect, it even makes it possible to ask the question why some people react so strongly to homophobia. Is there perhaps a more constructive response? Already we see that reframing the question consequentially enables conversation—conversation that, in principle, everyone can participate in, homophobes and antihomophobes alike. In reality of course, not everyone will engage. But at least everyone can receive a good faith invitation to.

The second consequence of this reframing is to make people aware of how much they don't know; it punctures their illusion of explanatory depth.[25] Asking someone to explain how something works often reveals that the person does not know. They discover they do not understand as well as they thought they did. In the example, it may be hard to explain why one person becomes a homophobe and somebody else who lives in a similar environment does not. We rarely know the life history and motivations of strangers, and our relative ignorance should make us more forgiving and less willing to pass judgment on them. Thus the attempt to generate a causal explanation often fails, and when it does, it can (and should) reduce hubris, make people feel less sure of their opinion, and open people up to alternative perspectives.

For most real-world issues, thinking consequentially involves thinking with more complexity. Confronting complexity can be quite profitable. As journalist Amanda Ripley put it, "When people encounter complexity, they become more curious, and less closed off to new information. They listen, in other words. There are many ways to complicate the narrative . . . But the main idea is to feature nuance, contradiction and ambiguity wherever you can find it."[26]

Nonetheless, I am not suggesting simply to think consequentially and not in terms of sacred values. If you have the time and ability to take on more complex questions, then you should. Frequently, though, people do not have the time, knowledge, or will to work their way through complex issues. If you come across a homophobe, you can only guess what their background is. Without learning a lot about the individual, there is no way for you to understand why they are the way they are with any certainty. Individual histories are too different from one another. So my advice is rarely simply to think with more complexity. Rather, it is to acknowledge the complexity of decisions and that we depend on others' knowledge and skills to make difficult decisions in order to decide with less hubris and more curiosity. What we are responsible for is to make sure the others that we rely on to support us are steeped in evidence, logic, and good faith, and not guided by their own self-interest.

12 Bottling Up Outrage

Sacred values have a lot in common with addictions. The addict is driven to act regardless of how bad or unpleasant the outcome is. Many heroin addicts wish fervently that they were off the drug as they watch their lives unravel while they take their next hit—an act that may not even lead them to feel good. Likewise, sacred values compel one to act according to their own dictates independent of the outcome. Neither addiction nor sacred values are about consequences.

Sacred values and addictions are similar not only in terms of what they lack. Both of them consist of action rules that get stamped in over time through reinforcement. Sacred values are rules that govern behavior—what one must do or must not do (thou shalt remember the Sabbath and thou shalt not kill). These are rules that can be articulated and debated. Associated with these behavioral rules are programs that govern our motor actions and visceral reactions. These action-reaction rules govern how sacred values get embodied and come to control us. If I live in a community that celebrates Super Bowl Sunday and I take part, then my community reinforces my own individual actions that celebrate the day. I am rewarded for watching the game by the camaraderie, excitement, music, and so on. Eventually, I don't really choose to watch the game on Super Bowl Sunday. Rather, I enact my Super Bowl Sunday script that has been stamped in through reinforcement without thinking about it. I watch the game that day. That's what I do. Similarly, once I have acquired the sacred value of not killing, then I get outraged when I see killing. I don't consciously choose to experience outrage; I simply react with outrage. By necessity, I have learned such a reaction. What has been the reinforcer guiding such learning? Presumably it is my conscience, a distillation of the sacred values of those around me. Sacred values are learned because they are reinforced by one's community. Either they are reinforced directly in the way that making one's bed is reinforced by a member of one's community—in my case, my mother. (Actually, neither of my parents gave a damn if I made my bed, and my bed remains unmade today.) Or they are reinforced indirectly through the internalization of community standards. For instance, there's a good chance that no one you know has killed anyone, and everyone reacts with horror at the thought.

Addiction also consists of action rules that get stamped in over time through reinforcement. Rules like "I've just finished dinner, so it is time for a cigarette" or "I'm sitting at my computer, so it is time to play a video game or engage social media." These are rules of action that are stamped in through repeated reinforcement.

Is it fair to conclude, given these similarities, that sacred values are a type of addiction? Would it be more fair to conclude that sacred values are simply a type of habit? In chapter 10, we saw that the only difference between an exercise habit and a substance abuse one is how we evaluate them. Most of us with an exercise habit are happy to keep it; most who abuse substances wish their lives were different. So the question of whether sacred values are better characterized as a habit or addiction amounts to the question of whether we want to keep our sacred values, or if we would prefer to shed them like an old layer of skin.

Changing Minds

Does anyone ever want to shed their sacred values? People do occasionally go through major life changes that bring them into new communities, either by choice or circumstance. Sometimes these new communities have their own cultures that clash with the sacred values the person acquired elsewhere. College students from conservative families can experience this when they move away to college and get embedded in liberal communities. Those who join the military sometimes discover an ethos in their new surroundings quite different from what they are used to. Migrants also can have this experience in a new country.[1] One way people deal with these kinds of tension is to shed old values and take on new ones.

But more often, people do not want to shed their sacred values so much as suspend their influence temporarily. Someone whose sacred values imply that only people of opposite sexes should get married might want to suppress their sacred value if their child informs them they are marrying someone of the same sex. And someone who is a strict pacifist may feel the need to hold their tongue and go along with the crowd if their community has just suffered a terrorist attack and reacts violently. In cases like these, a person might try to suppress a sacred value frame for a particular decision in favor of a consequentialist frame.

Even more often, people want others to adopt a consequentialist frame rather than a sacred values one. This occurs when the other person's sacred values clash with your own. You might be negotiating with someone from an out-group, or simply observing someone from an out-group negotiate or make a decision. You might want that person to adopt a consequentialist perspective because that would allow you to make an argument that is in your interest. Warring parties want the other side to see the battle in terms of potential gains and losses because war generally involves more loss than gain. If the other side can see the battle this way, it might stop fighting. You might even want the other side to take a consequentialist perspective because that will help it see what is in its own interests. Populist leaders are known to prey on sacred values. They use them on occasion to trick their constituents into behaving in ways that are antithetical to their own interests.

Sometimes people want others to adopt a consequentialist frame even though they share the other person's sacred values. We have already seen that sacred values may be absolute in principle but more flexible in practice. Few people will insist on fighting for their nation with the persistence of the Zealots if it becomes clear that doing so will result in your nation suffering a massacre. In that situation, you might try to convince your fellow combatants to minimize the group's losses by adopting a consequentialist frame. Similarly, you might want the person negotiating on your behalf to try to squeeze as much as possible out of the other side rather than to fail outright by adopting an absolutist position based on sacred values.

It is under conditions like these—when people want to let go of or temporarily suppress their sacred values—that it is useful to equate sacred values with addictions. The literature on addiction provides some insights into how to help people free themselves of a sacred value.

How to Cure a Sacred Value

In chapter 10, we encountered the visceral theory of addiction.[2] It attributes addiction to strong cravings and posits that cravings are induced by associations to environmental cues. It proposes that the main impediment to quitting an addiction is the problem of craving-induced relapse (and not the desire to avoid withdrawal symptoms). We fail to appreciate that we are caught in an addiction because most of the time we do not experience craving. More important, when we are not in the throes of craving, we fail to appreciate how strong our cravings are and how they tend to overwhelm us. Those who quit smoking for a while fail to appreciate how powerful their future cravings will be as well as how their cravings will take over and cause them to think obsessively about how much they want a cigarette, so that they can no longer think about anything else, including other people's needs or the future.

We fail to appreciate the power of craving because our imaginations cannot reproduce the force that cravings have in reality. Consider an analogy. If I ask you to imagine the face of the current president of the United States, most of you can do it to a reasonable extent. Visual stimuli can be imagined and also at least partially reconstructed in memory. The same is true of auditory images. Most people can retrieve their favorite song from memory and play it in their head. They are not playing the actual song but instead a rendition of it that is unmistakably like the actual song. Now try to do the same with smell. Can you imagine the smell of a rose? Or garbage? You can recognize those smells certainly, but the vast majority of people find it hard to generate smells without the stimulus present.

Craving is more like smell than it is like vision or hearing. It is hard, if not impossible, to imagine what it is like to crave without being in a state of craving. And a state of craving can take us over. When we are really hungry or thirsty, our hunger or thirst is all we can think about. We will take actions that we are not proud of to slake our thirst or relieve our hunger. Addicts will beg, borrow, steal, and worse to satisfy their craving.

According to the visceral theory, the very ease of stopping an addiction in the short run may exacerbate the difficulty of stopping in the long run. You may be able to overcome your cravings for a period, but doing so can give you the impression that you have the addiction under control and sufficient willpower to overcome the addiction, precisely because it is hard to imagine how powerful cravings can get. And the belief that you are in control gives you permission to succumb to the addiction later.

To eliminate craving, the visceral theory prescribes removing the cues that elicit cravings. If you smoke in your dining room after dinner, then at the end of dinner you should leave the dining room. If you gamble whenever you are with your buddies Pete and Rose, then stay away from Pete and Rose. If you cannot stop eating cashews when they are on the table in front of you, then take them off the table.

Sometimes we cannot control whether cues are present. It might be rude to leave the dining room immediately after dinner, Pete and Rose might be too much a part of our lives to stay away from, and you might not be the one who gets to decide whether or not the cashews are on the table. In that case, you are stuck doing what Odysseus did to avoid crashing on the rocks when the Sirens called. He stuffed his sailors' ears with beeswax so they could not hear the Sirens or his commands, and had them tie him to the mast so he couldn't jump in the water. That way, he was able to hear the infinitely alluring song of the Sirens without being drawn to his death. As an addict, you can tie yourself to the mast by making sure not to have cigarettes around or the means to get them after dinner, or by hanging out with Pete and Rose only when you have no money and so can't gamble. Taking these steps requires having an appreciation for the power of a future craving even when you are not experiencing it.

If sacred values are addictions, then we should be able to apply this lesson to sacred values. That requires a clear understanding of what we are addicted to when we have a sacred value and, in particular, what cues elicit the addictive reaction and what sacred values cause us to crave. If my body has internalized the belief that murder is wrong, then I desire that people do not kill one another. But this does not really qualify as a craving. Rarely are people obsessed with a desire for people not to murder one another. Instead, the craving is for justice when I do see a murder. Then, I experience a reaction of outrage or anger that can be hard to ignore. Similarly, when abortion foes hear about an abortion, they are outraged. And when progressives come across a statement that smells of racism or misogyny, they are outraged.

It is violation of the sacred value that evokes a craving—for justice or revenge, or whatever it takes to make the world right. That is why dialogue is so difficult when issues are framed in terms of sacred values because the possibility of a violation evokes a craving that focuses the individual on their own need, engulfing their field and inhibiting their ability to see other sides of the issue.

So the simple lesson is that if you want to avoid a sacred values frame, you need to avoid being outraged. Avoid cable news whose business model is to outrage viewers. Avoid politicians whose strategy is to sow fear and outrage by talking about all the evils of their opponents. Avoid podcasts and radio shows that consist of reporting a litany of outrageous behaviors by the people they hate. And avoid hanging out with people who insist on telling stories designed to upset you. The reason is that it is hard to put outrage back in the bottle. Once our ire has been raised, we crave some kind of rectification, some correction or at least redress for the wicked or vile act that has been performed. The sense of righteousness does not let go easily. It behaves the same way as an addiction.

On Ideological Health

Choice is hard. Modern life offers no shortage of choices. Therefore modern life is hard. When decisions require that individual opinions are aggregated into some kind of consensus, choice is even harder. At the time of this writing (October 2023), the US House of Representatives does not have a speaker because the dominant Republican Party cannot reach consensus on one. As a result, Congress has been left impotent, unable to make decisions.

More broadly, there are societal needs that everyone recognizes but society cannot act on because we cannot agree on the right solution. Everyone agrees that internet speech requires at least a modicum of regulation. At minimum, people should not have the freedom to outright lie about others in ways that could produce real harm. Even the most extreme proponents of freedom of speech want to prevent lies about themselves. But nobody has figured out how to navigate the complex knot of issues that arise when you try to limit speech. Should everyone's speech be limited, including that of political leaders? Exactly what speech should be restricted? Who gets to decide? How should violators be punished? There have been valiant attempts to solve these problems, but no solution with broad agreement is on the horizon.

One reason collective choice problems are hard is because communicating about them requires a shared narrative. Different communities sometimes just do not and cannot agree on a narrative. They sometimes disagree about the facts. For instance, was there or was there not evidence of weapons of mass destruction in Iraq before the US invasion in 2003? More often, narratives disagree about the interpretation of the fact. Was the bombing of Al-Ahli Baptist Hospital in Gaza on October 17, 2023, for example, done by Israel, as Hamas claims, or a failed rocket launch by Islamic Jihad, as Israel claims?

We have seen that integral emotions are essential for sound decision making, but they also introduce bias. For instance, after airplanes were hijacked in the 9/11 attacks in 2001, fear of flying led to more driving. Yet the base rates for death from driving are higher than those from flying the equivalent mileage. As a result, the reluctance to fly due to fear for the three months following the 9/11 attacks led to more deaths among travelers.[3] Fear of dying increased the likelihood of dying. Integral emotions that arise from violations of sacred values, like anger and outrage, are partly responsible for decisions that end in tragedy like war and devastation. Competing sacred values stir up conflict by giving their respective communities a sense of common purpose.

So there is reason to reduce our reliance on sacred values when our aim is to reach consensus. One approach is temporary—to reframe issues that we tend to think about in sacred values terms in consequentialist terms. The last chapter pointed out that one can do that through causal explanation. So far, such an intervention has only been tried at the individual level. Individuals have been asked for causal explanations. While the question does reduce individuals' extremity, it is a small effect that is not robust.[4] We need to try an intervention like this at a collective level. We need to change social norms to put causal explanations at the heart of political discourse.

But such temporary reframings of issues are not cure-alls. Serious social change requires approaches that influence habits of thought, not merely perspectives held in the short term. At the individual level, therapy serves as a good model for this. Indeed, there are therapeutic methods that take cultural values as a starting point. Value-based therapy rests on the assumption that each client must be understood within their personal frame of values that emanate from their cultural experience.[5] According to psychological counselor and value-based therapist Inge Missmahl, who has worked with many migrants from Afghanistan, the stress from exposure to violence, loss and grief, and difficult life transitions as well as issues of honor, shame, and abuse leads to both somatic and mental health symptoms. Migrants can suffer severe stress when they are surrounded by people doing things unheard of in their culture. We all experience this to some extent due to rapid transitions occurring in society.

Missmahl describes value-based therapy this way: "Together with the client, she/he explores the meaning of the symptom in the given cultural and social context of the client based on the client's own perception. The counsellor is guided by the following questions: What is the person expressing with this symptom? What is the psychosocial stressor which is related to the symptom and what is their relationship to each other? . . . The aim is to reach a shared understanding of the inner situation of the client."[6] Sacred values are deeply embedded in a person's cultural identity. Change is thus highly disruptive. It reverberates through a person's understanding of themselves, their personal relationships, and their relation to work and other institutions. This makes uprooting sacred values hard and even a potential risk.

Change at the societal level requires the development of new norms—norms that foster consequentialist talk and recognize the inadequacy of sacred values talk. Changing norms means putting new standards of conversation into place. Those standards need to give people the tools to challenge one another productively. Challenging one another has come to be interpreted as a hostile, conflictual act in many corners of America in recent years according to First Amendment lawyer Greg Lukianoff and social psychologist Jonathan Haidt. People in the United States have become more and more conflict averse within their bubbles as they strive to protect their children from both physical harm and emotional trauma. One example they offer is the prioritization of safety. Safety has become a sacred value leading to an unwillingness to trade off pleasure, learning, or anything else with the perception that they and their children are safe. As a result, according to Lukianoff and Haidt, we are living in a culture of "safetyism" where safety "trumps everything else, no matter how unlikely or trivial the potential danger."[7] The unfortunate consequence of such safetyism is that when conflict and threat do arise, as they inevitably do, people have not had the experience necessary to learn how to deal with them.

One effect of this avoidance of conflict is that it has become hard to tell each other when our arguments and justifications for our positions are inadequate. But if all we do to explain our position is appeal to a sacred value, that is inadequate, and we need to be able to say so. A serious justification must also consider the consequences of the position. People do not generally like being told that their argument does not meet standards, so it is important that this kind of conflict be understood as constructive and acceptable. Our discourse standards need to also engender an acceptance of wildly varying points of view. We have to be able to challenge one another without dismissing one another's perspective. Walking this fine line is difficult, but teachers, parents, and other leaders do it all the time, so it must be doable.

Looking at events through our sacred values helps us avoid the harrowing complexity of the real world that we live in. This reliance on sacred values is part of the causal chain leading to the culture wars in the United States today, although the tendency is not unique to the modern US. It describes previous historical epochs in the US like the revolutionary period, Civil War era, and 1960s with its civil rights movement, Vietnam War, and generation gap. And it has happened in other countries (the Zealots two thousand years ago, Crusades, Reformation, etc.). In each case, strife has complex political and economic causes—battles over power along with competing interests. In each case, sacred values became the medium that enabled strife to turn into polarization, into groups battling in their own corners unwilling and unable to see a competing perspective.

How can we induce people to see beyond their sacred values to do the hard work of trying to understand the short- and long-term consequences of their beliefs, attitudes, and actions? By demonstrating that the hard work pays off. In the end, it is consequences that matter, not regulating outrage. Climate change will be solved if and only if we figure out what causes it and how to stop it, not by expressions of outrage at climate change deniers. The conflicts in the Middle East will only be solved when peace offers all parties more rewards than fighting, not by simplistic characterizations of warring factions as good or evil.

It is true that we must simplify to communicate, especially when the audience is large and varied. Sacred values provide a means to accomplish that. It is also true that leadership requires motivating people. Sacred values provide a means to accomplish that too. But a leadership that has not been fully informed by a disinterested consequentialist analysis is not only going to fail to achieve the best results, it is violating its sacred duty to its followers.

Acknowledgments

First and foremost, I must thank the person who read my drafts carefully and turned my wordy and laborious prose into text that is lighter and more readable, my daughter Leila Sloman. She also removed some of my weaker jokes (though I put a few back when she wasn't looking). I received valuable feedback on the manuscript from my wife, Linda Covington, and my mother, Valerie Sloman, as well as Laura Niemi, Michael Shiner, Mugur Geana, Michael Paul, and Alex Szebenyi.

Some of the ideas were inspired by conversations with my other daughter, Sabina Sloman, and with my friends Phil Fernbach, Dan Bartels, and Phil Leis. I have benefited immensely from countless conversations on the topics of the book with associates of my lab, especially Hyosoek Kim, Victoria Halewicz, Almos Molnar, Semir Tatlidil, Tao Burga, and Mimi Jessop.

Notes

Chapter 1

1. Grayzel, S. (1960). *A history of the Jews, from the Babylonian exile to the establishment of Israel*. Jewish Publication Society of America.

2. Tetlock, P. E., Kristel, O. V., Elson, S. B., Green, M. C., & Lerner, J. S. (2000). The psychology of the unthinkable: Taboo trade-offs, forbidden base rates, and heretical counterfactuals. *Journal of Personality and Social Psychology, 78*(5), 853–870.

3. Rozenblit, L., & Keil, F. C. (2002). The misunderstood limits of folk science: An illusion of explanatory depth. *Cognitive Science, 26*, 521–562; Sloman, S. A., & Vives, M. L. (2022). Is political extremism supported by an illusion of understanding? *Cognition, 225*, 105146; Zeveney, A. S., & Marsh, J. K. (2016). The illusion of explanatory depth in a misunderstood field: The IOED in mental disorders. In A. Pagafragou, D. Grodner, D. Mirman, & J. C. Trueswell (Eds.), *Proceedings of the 38th annual conference of the cognitive science society* (pp. 1020–1025). Cognitive Science Society.

4. Quoted in Williams, Z. (2016, October 28). Never give a straight answer: How I learned to talk like a politician. *Guardian*. https://www.theguardian.com/politics/2016/oct/28/never-give-a-straight-answer-zoe-williams-learns-to-talk-like-a-politician.

5. Greene, J. D., Sommerville, R. B., Nystrom, L. E., Darley, J. M., & Cohen, J. D. (2001). An fMRI investigation of emotional engagement in moral judgment. *Science, 293*(5537), 2105–2108. See also Sloman, S. A. (1996). The empirical case for two systems of reasoning. *Psychological Bulletin, 119*(1), 3.

6. See Sloman, S., & Fernbach, P. (2017). *The knowledge illusion: Why we never think alone*. Riverhead Books.

7. Matt. 7:12.

Chapter 2

1. For instance, one of the *Merriam-Webster* dictionary definitions is "worthy of religious veneration." *Merriam-Webster.* (n.d.). Sacred. Retrieved August 27, 2024, from https://www.merriam-webster.com/dictionary/sacred.

2. Handfield, T. (2020). The coevolution of sacred value and religion. *Religion, Brain and Behavior, 10*(3), 252–271.

3. How religious are Americans? (2024, March 29). Gallup. https://news.gallup.com/poll/358364/religious-americans.aspx.

4. Fodor, J. A. (1983). *The modularity of mind* (p. 37). MIT Press.

5. Baron, J., & Spranca, M. (1997). Protected values. *Organizational Behavior and Human Decision Processes, 70*(1), 3.

6. Marietta, M. (2008). From my cold, dead hands: Democratic consequences of sacred rhetoric. *Journal of Politics, 70*(3), 767–779.

7. Baron, J., & Leshner, S. (2000). How serious are expressions of protected values? *Journal of Experimental Psychology: Applied, 6*(3), 183.

8. Kreps, T. A., & Monin, B. (2014). Core values versus common sense: Consequentialist views appear less rooted in morality. *Personality and Social Psychology Bulletin, 40*(11), 1529–1542.

9. Lichtenstein, S., Gregory, R., & Irwin, J. (2007). What's bad is easy: Taboo values, affect, and cognition. *Judgment and Decision Making, 2,* 169–188.

10. Crockett, M. J. (2017). Moral outrage in the digital age. *Nature Human Behaviour, 1,* 769–771.

11. For supporting data, see Tetlock, P. E., Kristel, O. V., Elson, S. B., Green, M. C., & Lerner, J. S. (2000). The psychology of the unthinkable: Taboo trade-offs, forbidden base rates, and heretical counterfactuals. *Journal of Personality and Social Psychology, 78*(5), 853–870.

12. Perlstein, R. (2020, August 17). An interview with "Playboy" magazine nearly torpedoed Jimmy Carter's presidential campaign. *Playboy.* https://www.smithsonianmag.com/history/interview-playboy-magazine-nearly-torpedoed-jimmy-carters-presidential-campaign-180975576/.

13. A point made in Tetlock et al. (2000).

14. Klein, E. (2020). *Why we're polarized.* Simon and Schuster.

15. Baron & Leshner (2000).

16. Lichtenstein & Irwin (2007); Pincus M., LaViers L., Prietula M. M. J., & Berns G. G. (2014). The conforming brain and deontological resolve. *PLoS ONE, 9*, e106061. doi:10.1371/journal.pone.0106061.

17. Ritov, I., & Baron, J. (1999). Protected values and omission bias. *Organizational Behavior and Human Decision Processes, 79*(2), 83.

18. Baron & Spranca (1997); Tanner, C., Medin, D. L., & Iliev, R. (2008). Influence of deontological versus consequentialist orientations on act choices and framing effects: When principles are more important than consequences. *European Journal of Social Psychology, 38*(5), 757–769.

19. Ritov & Baron (1999).

20. Baron & Leshner (2000).

21. Ritov & Baron (1999).

22. Bartels, D. M., & Medin, D. L. (2007). Are morally motivated decision makers insensitive to the consequences of their choices? *Psychological Science, 18*(1), 24–28.

23. Tetlock et al. (2000).

24. Skitka, L. J., & Houston, D. A. (2001). When due process is of no consequence: Moral mandates and presumed defendant guilt or innocence. *Social Justice Research, 14*, 305–326.

25. Atran, S., & Axelrod, R. (2008). Reframing sacred values. *Negotiation Journal, 24*(3), 221–246.

26. Haidt, J. (2012). *The righteous mind: Why good people are divided by politics and religion*. Vintage.

27. Atari, M., Haidt, J., Graham, J., Koleva, S., Stevens, S. T., & Dehghani, M. (2023). Morality beyond the WEIRD: How the nomological network of morality varies across cultures. *Journal of Personality and Social Psychology, 125*(5), 1157–1188.

28. Sartre, J. P. (1944). Paris alive: The republic of silence. *Atlantic Monthly, 174*(6), 39–40.

29. Can psychopaths have friends & intimate relationships? (n.d.). Psychopaths in Life. Retrieved September 17, 2024, from https://psychopathsinlife.com/can-psychopaths-have-friends-relationships/.

30. Patil, I. (2015). Trait psychopathy and utilitarian moral judgement: The mediating role of action aversion. *Journal of Cognitive Psychology, 27*(3), 349–366.

31. Everett, J. A. C., Pizarro, D. A., & Crockett, M. J. (2016). Inference of trustworthiness from intuitive moral judgments. *Journal of Experimental Psychology: General, 145*(6), 772–787.

32. Ford Pinto. (n.d.). Wikipedia. Retrieved September 17, 2024, from https://en.wikipedia.org/wiki/Ford_Pinto.

33. Who we are. (n.d.). Republican National Committee. Retrieved September 17, 2024, from https://gop.com/about-our-party/; Party platform: The Democratic platform. (n.d.). Democratic National Committee. Retrieved September 17, 2024, from https://democrats.org/where-we-stand/party-platform/.

34. Quoted in Borowitz, A. (2022). *Profiles in ignorance: How America's politicians got dumb and dumber.* Simon and Schuster.

35. Forbes, J. D. (2001). Indigenous Americans: Spirituality and ecos. *Daedalus, 130*(4), 283–300.

36. Wertsch, J. V. (2008). Collective memory and narrative templates. *Social Research: An International Quarterly, 75*(1), 133–156.

37. Sloman, S., & Fernbach, P. (2017). *The knowledge illusion: Why we never think alone.* Riverside Press.

38. Schwartz, B. (1987). *The battle for human nature: Science, morality and modern life.* W. W. Norton and Company.

39. Haidt, J., & Joseph, C. (2004). Intuitive ethics: How innately prepared intuitions generate culturally variable virtues. *Daedalus, 133*(4), 55–66.

40. Graham, J., Atari, M., Dehghani, M., & Haidt, J. (2023). Puritanism needs purity, and moral psychology needs pluralism. *Behavioral and Brain Sciences, 46*(e293), 42–43.

41. DeScioli, P., & Kurzban, R. (2023). Moralistic punishment is not for cooperation. *Behavioral and Brain Sciences, 46*(e293), 34–35.

42. Fitouchi, L., André J.-B., & Baumard N. (2023). Moral disciplining: The cognitive and evolutionary foundations of puritanical morality. *Behavioral and Brain Sciences, 46*(e293), 1–74.

Chapter 3

1. Levin, M., & Pinkerson, D. (Directors). (2000). *Soldiers in the army of god* [Film]. Off Line Entertainment Group.

2. Paul Jennings Hill. (n.d.). Wikipedia. Retrieved September 23, 2024, from https://en.wikipedia.org/wiki/Paul_Jennings_Hill.

3. Marietta, M. (2008). From my cold, dead hands: Democratic consequences of sacred rhetoric. *Journal of Politics, 70*(3), 768.

4. Mintz, S. (2021, June 21). Decolonizing the academy. *Higher Ed Gamma.* https://www.insidehighered.com/blogs/higher-ed-gamma/decolonizing-academy.

5. Kruglanski, A. W., Szumowska, E., Kopetz, C. H., Vallerand, R. J., & Pierro, A. (2021). On the psychology of extremism: How motivational imbalance breeds intemperance. *Psychological Review, 128*(2), 264.

6. Coleman, P. (2021). *The way out: How to overcome toxic polarization.* Columbia University Press.

7. Marietta (2008), pp. 768–769.

8. See, for example, Halbrook, S. (2023, June 6). Second Amendment roundup: Looking for historical analogues in all the wrong places. *Volokh Conspiracy.* https://reason.com/volokh/2023/06/06/second-amendment-roundup-looking-for-historical-analogues-in-all-the-wrong-places/.

9. Gramlich, J. (2023, April 26). What the data says about gun deaths in the U.S. Pew Research Center. https://www.pewresearch.org/fact-tank/2022/02/03/what-the-data-says-about-gun-deaths-in-the-u-s/.

10. Gillespie, N. (2023, April 5). Do "more guns lead to more deaths"? [Video]. *Reason.* https://reason.com/video/2023/04/05/do-more-guns-lead-to-more-deaths/.

11. Skitka, L. J., Hanson, B. E., Morgan, G. S., & Wisneski, D. C. (2021). The psychology of moral conviction. *Annual Review of Psychology, 72*, 347–366.

12. Ryan, T. J. (2017). No compromise: Political consequences of moralized attitudes. *American Journal of Political Science, 61*(2), 413.

13. Kodapanakkal, R. I., Brandt, M. J., Kogler, C., & van Beest, I. (2022). Moral frames are persuasive and moralize attitudes; nonmoral frames are persuasive and de-moralize attitudes. *Psychological Science, 33*(3), 433–449.

14. Delton, A. W., DeScioli, P., & Ryan, T. J. (2020). Moral obstinacy in political negotiations. *Political Psychology, 41*(1), 3–20.

15. Atari, M., Haidt, J., Graham, J., Koleva, S., Stevens, S. T., & Dehghani, M. (2023). Morality beyond the WEIRD: How the nomological network of morality varies across cultures. *Journal of Personality and Social Psychology, 125*(5), 1157–1188.

16. Marietta (2008).

17. Sartre, J.-P. (1944). Paris alive: The republic of silence. *Atlantic Monthly, 174*(6), 39.

18. Neely, P. (2013, May 3). Jamestown colonists resorted to cannibalism. *National Geographic.* https://www.nationalgeographic.com/science/article/130501-jamestown-cannibalism-archeology-science.

19. Bradatan, C. (2015). *Dying for ideas: The dangerous lives of the philosophers.* Bloomsbury Publishing.

20. Atran, S. (2010). *Talking to the enemy: Violent extremism, sacred values, and what it means to be human*. Penguin.

21. Atran (2010), p. 178.

22. Dehghani, M., Iliev, R., Sachdeva, S., Atran, S., Ginges, J., & Medin, D. (2009). Emerging sacred values: Iran's nuclear program. *Judgment and Decision Making, 4*(7), 530–533.

23. Greene, J. (2014). *Moral tribes: Emotion, reason, and the gap between us and them*. Penguin.

24. Greene (2014), p. 302.

Chapter 4

1. What is effective altruism? (n.d.). Effective Altruism. Retrieved September 23, 2024, from https://www.effectivealtruism.org/articles/introduction-to-effective-altruism.

2. The originals came from Singer, P. (1972). Famine, affluence and morality. *Philosophy and Public Affairs, 1*, 229–243. This version was lifted from Greene, J. (2009). *What's next: Dispatches on the future of science* (pp. 104–115). Vintage.

3. Mill, J. S. (1863). *Utilitarianism* (p. 14). London: Parker, Son, and Bourn.

4. Mill (1863), p. 257.

5. Kahane, G., Everett, J. A., Earp, B. D., Caviola, L., Faber, N. S., Crockett, M. J., & Savulescu, J. (2018). Beyond sacrificial harm: A two-dimensional model of utilitarian psychology. *Psychological Review, 125*(2), 131.

6. Singer, P. (1979). *Practical ethics*. Cambridge University Press; Singer, P. (2015). *The most good you can do: How effective altruism is changing ideas about living ethically*. Yale University Press.

7. Berridge, K. C. (2009). Wanting and liking: Observations from the neuroscience and psychology laboratory. *Inquiry, 52*(4), 378–398.

8. Joo, M., Liu, W., & Wilbur, K. C. (2020). Divergent temporal courses for liking versus wanting in response to persuasion. *Emotion, 20*(2), 261.

9. This is reminiscent of the classic mere exposure effect. See Zajonc, R. B. (1950). Preferences need no inferences. *American Psychologist, 35*(2), 25.

10. Schkade, D. A., & Kahneman, D. (1998). Does living in California make people happy? A focusing illusion in judgments of life satisfaction. *Psychological Science, 9*(5), 340–346.

11. Frankl, V. E. (1964). *Ein Psycholog erlebt das Konzentrationslager* [*Man's search for meaning: An introduction to logotherapy*] (I. Lasch, Trans.) (Rev. ed.). Hodder & Stoughton.

12. Kruglanski, A. W., Gelfand, M. J., Bélanger, J. J., Sheveland, A., Hetiarachchi, M., & Gunaratna, R. (2014). The psychology of radicalization and deradicalization: How significance quest impacts violent extremism. *Political Psychology, 35,* 69–93.

13. Kruglanski, A. W., Molinario, E., Jasko, K., Webber, D., Leander, N. P., & Pierro, A. (2022). Significance-quest theory. *Perspectives on Psychological Science, 17*(4), 1050–1071.

Chapter 5

1. See Tulving, E., & Thomson, D. M. (1973). Encoding specificity and retrieval processes in episodic memory. *Psychological Review, 80*(5), 352.

2. Berliner, P. F. (1994). *Thinking in jazz: The infinite art of improvisation* (p. 102). University of Chicago Press.

3. Payne, J. W., Bettman, J. R., & Johnson, E. J. (1983). *The adaptive decision maker.* Cambridge University Press.

4. Tversky, A., & Kahneman, D. (1983). Extensional versus intuitive reasoning: The conjunction fallacy in probability judgment. *Psychological Review, 90*(4), 293.

5. Sloman, S. A., Over, D., & Slovak, L. (2003). Frequency illusions and other fallacies. *Organizational Behavior and Human Decision Processes, 91*(2), 296–309.

6. Fiedler, K. (1988). The dependence of the conjunction fallacy on subtle linguistic factors. *Psychological Research, 50*(2), 125. He included other options for students to rank as well, but those are irrelevant for our purposes.

7. Gigerenzer, G., & Goldstein, D. G. (2011). The recognition heuristic: A decade of research. *Judgment and Decision Making, 6*(1), 100–121.

8. Barsalou, L. W. (1991). Deriving categories to achieve goals. In G. H. Bower (Ed.), *Psychology of learning and motivation* (Vol. 27, pp. 1–64). Academic Press.

9. Blanton, H., Jaccard, J., Klick, J., Mellers, B., Mitchell, G., & Tetlock, P. E. (2009). Strong claims and weak evidence: Reassessing the predictive validity of the IAT. *Journal of Applied Psychology, 94*(3), 567.

10. ninofreno. (2023, February 3). "Human, all too human": ChatGPT and the conjunction fallacy. *Durissima Coquere.* https://ninofreno.com/2023/02/03/human-all-too-human-chatgpt-and-the-conjunction-fallacy/.

11. Hagendorff, T., & Fabi, S. (2023). *Human-like intuitive behavior and reasoning biases emerged in language models—and disappeared in GPT-4.* arXiv. arXiv:2306.07622.

12. Gilovich, T., Vallone, R., & Tversky, A. (1985). The hot hand in basketball: On the misperception of random sequences. *Cognitive Psychology, 17*(3), 295–314.

13. Metzger, M. A. (1985). Biases in betting: An application of laboratory findings. *Psychological Reports, 56*(3), 883–888; Keren, G. B., & Wagenaar, W. A. (1985). On the psychology of playing blackjack: Normative and descriptive considerations with implications for decision theory. *Journal of Experimental Psychology: General, 114*(2), 133; Suetens, S., Galbo-Jørgensen, C. B., & Tyran, J. R. (2016). Predicting lotto numbers: A natural experiment on the gambler's fallacy and the hot-hand fallacy. *Journal of the European Economic Association, 14*(3), 584–607; Xu, J., & Harvey, N. (2014). Carry on winning: The gamblers' fallacy creates hot hand effects in online gambling. *Cognition, 131*(2), 173–180.

14. Klayman, J. (1995). Varieties of confirmation bias. *Psychology of Learning and Motivation, 32*, 385–418.

15. Troutman, C. M., & Shanteau, J. (1977). Inferences based on nondiagnostic information. *Organizational Behavior and Human Performance, 19*(1), 43–55.

16. Lord, C. G., Ross, L., & Lepper, M. R. (1979). Biased assimilation and attitude polarization: The effects of prior theories on subsequently considered evidence. *Journal of Personality and Social Psychology, 37*(11), 209; Taber, C. S., & Lodge, M. (2006). Motivated skepticism in the evaluation of political beliefs. *American Journal of Political Science, 50*(3), 755–769.

17. Dick Cheney's suite demands. (n.d.). *Smoking Gun*. Retrieved September 26, 2024, from https://www.thesmokinggun.com/file/dick-cheneys-suite-demands.

18. Adams, J. S. (1961). Reduction of cognitive dissonance by seeking consonant information. *Journal of Abnormal and Social Psychology, 62*(1), 74–78.

19. Hart, W., Albarracín, D., Eagly, A. H., Brechan, I., Lindberg, M. J., & Merrill, L. (2009). Feeling validated versus being correct: A meta-analysis of selective exposure to information. *Psychological Bulletin, 135*(4), 555.

20. Del Vicario, M., Bessi, A., Zollo, F., Petroni, F., Scala, A., Caldarelli, G., . . . & Quattrociocchi, W. (2016). The spreading of misinformation online. *Proceedings of the National Academy of Sciences, 113*(3), 554–559.

21. Stangor, C., & McMillan, D. (1992). Memory for expectancy-congruent and expectancy-incongruent information: A review of the social and social developmental literatures. *Psychological Bulletin, 111*(1), 42–61.

22. Rasyid, S., Kunda, Z., & Fong, G. T. (1990). Motivated recruitment of autobiographical memories. *Journal of Personality and Social Psychology, 59*(2), 229–241.

23. Mercier, H., & Sperber, D. (Eds.). (2017). *The enigma of reason*. Harvard University Press; Page, S. E. (2008). *The difference: How the power of diversity creates better groups, firms, schools, and societies* (new ed.). Princeton University Press; Popper, K. (1962). *Conjectures and refutations: The growth of scientific knowledge*. Basic Books.

24. Sloman, S., & Fernbach, P. (2017). *The knowledge illusion: Why we never think alone*. Riverside Press.

25. Sloman, S. A. (2005). *Causal models: How people think about the world and its alternatives*. Oxford University Press.

Chapter 6

1. Lewis, D. (1973). Causation. *Journal of Philosophy, 70*(17), 556–567; Gerstenberg, T., Goodman, N. D., Lagnado, D. A., & Tenenbaum, J. B. (2021). A counterfactual simulation model of causal judgments for physical events. *Psychological Review, 128*(5), 936.

2. Woodward, J. F. (2009). Agency and interventionist theories. In H. Beebee, C. Hitchcock, & P. Menzies (Eds.), *The Oxford handbook of causation* (pp. 234–263). Oxford University Press.

3. Dowe, P. (2000). *Physical causation*. Cambridge University Press.

4. Walsh, C. R., & Sloman, S. A. (2011). The meaning of cause and prevent: The role of causal mechanism. *Mind & Language, 26*(1), 21–52; Wolff, P. (2007). Representing causation. *Journal of Experimental Psychology: General, 136*, 82–111; Lombrozo, T. (2010). Causal–explanatory pluralism: How intentions, functions, and mechanisms influence causal ascriptions. *Cognitive Psychology, 61*(4), 303–332.

5. Sloman, S. A., & Hagmayer, Y. (2006). The causal psycho-logic of choice. *Trends in Cognitive Sciences, 10*(9), 407–412.

6. This is the logic of intervention spelled out in Pearl, J. (2000). *Models, reasoning and inference*. Cambridge University Press. I made intensive use of it in a book on the psychology of causal reasoning. See Sloman, S. (2005). *Causal models: How people think about the world and its alternatives*. Oxford University Press.

7. Pearl, J. (2014). Comment: Understanding Simpson's paradox. *American Statistician, 68*(1), 8–13.

8. Tversky, A., & Kahneman, D. (1982). Causal schemata in judgements under uncertainty. In D. Kahneman, P. Slovic, & A. Tversky (Eds.), *Judgment under uncertainty: Heuristics and biases* (pp. 117–128). Cambridge University Press.

9. Kahneman, D., & Tversky, A. (1982). The simulation heuristic. In D. Kahneman, P. Slovic, & A. Tversky (Eds.), *Judgment under uncertainty: Heuristics and biases* (pp. 201–208). Cambridge University Press.

10. Kahneman & Tversky (1982), p. 203.

11. Klein, G., Calderwood, R., & Clinton-Cirocco, A. (2010). Rapid decision making on the fire ground: The original study plus a postscript. *Journal of Cognitive Engineering and Decision Making, 4*(3), 194–195.

12. Langer, E. J. (1975). The illusion of control. *Journal of Personality and Social Psychology, 32*(2), 311–328.

13. Kahneman, D., & Renshon, J. (2009, October 13). Why hawks win. *Foreign Policy*. https://foreignpolicy.com/2009/10/13/why-hawks-win/.

14. Adelman, K. (2002, February 12). Cakewalk in Iraq. *Washington Post*. https://www.washingtonpost.com/archive/opinions/2002/02/13/cakewalk-in-iraq/cf09301c-c6c4-4f2e-8268-7c93017f5e93/.

15. Kahneman, D., & Renshon, J. (2007). Why hawks win. *Foreign policy* (158), 37.

16. Kahneman, D. (2011). *Thinking, fast and slow*. Macmillan.

17. Ballard, J. (2019, September 13). Do Americans really believe Friday the 13th is unlucky? YouGov. https://today.yougov.com/topics/society/articles-reports/2019/09/13/friday-the-13th-unlucky-poll-survey.

18. Risen, J. L. (2016). Believing what we do not believe: Acquiescence to superstitious beliefs and other powerful intuitions. *Psychological Review, 123*(2), 182.

19. Michotte, A. (1946). *La perception de la causalité* [*The perception of causality*] (R. Miles & E. Miles, Trans.). Editions de l'institut supérieur de Philosophie.

20. Chapman, L. J., & Chapman, J. P. (1969). Illusory correlation as an obstacle to the use of valid psychodiagnostic signs. *Journal of Abnormal Psychology, 74*(3), 271–280.

21. Campbell, D. E., & Beets, J. L. (1978). Lunacy and the moon. *Psychological Bulletin, 85*(5), 1123–1129; Rotton, J., & Kelly, I. W. (1985). Much ado about the full moon: A meta-analysis of lunar-lunacy research. *Psychological Bulletin, 97*(2), 286–306.

22. DeStefano, F., & Shimabukuro, T. T. (2019). The MMR vaccine and autism. *Annual Review of Virology, 6*, 585–600.

23. U.S. economic performance under Democratic and Republican presidents. (n.d.). Wikipedia. Retrieved September 26, 2024, from https://en.wikipedia.org/wiki/U.S._economic_performance_by_presidential_party.

24. Paluck, E. L., & Green, D. P. (2009). Prejudice reduction: What works? A review and assessment of research and practice. *Annual Review of Psychology, 60*, 339–367.

25. Drunk driving. (n.d.). National Highway Traffic Safety Administration. Retrieved September 26, 2024, from https://www.nhtsa.gov/risky-driving/drunk-driving.

26. Coudevylle, G. R., Boulley-Escriva, G., Finez, L., Eugène, K., & Robin, N. (2020). An experimental investigation of claimed self-handicapping strategies across motivational climates based on achievement goal and self-determination theories. *Educational Psychology, 40*(8), 1002–1021.

27. McCrea, S. M. (2008). Self-handicapping, excuse making, and counterfactual thinking: Consequences for self-esteem and future motivation. *Journal of Personality and Social Psychology, 95*(2), 274.

28. Jones, E. E., & Berglas, S. (1978). Control of attributions about the self through self-handicapping strategies: The appeal of alcohol and the role of underachievement. *Personality and Social Psychology Bulletin, 4*(2), 200–206.

29. Cherry, K. (2023, October 4). What is self-handicapping in psychology? Verywell Mind. https://www.verywellmind.com/self-handicapping-protecting-the-ego-at-a-cost-4125125.

30. Fernbach, P. M., Hagmayer, Y., & Sloman, S. A. (2014). Effort denial in self-deception. *Organizational Behavior and Human Decision Processes, 123*(1), 1–8.

31. Glymour, C. (2003). Learning, prediction and causal Bayes nets. *Trends in Cognitive Sciences, 7*(1), 43–48.

32. This is the topic of my book with Phil Fernbach, *The Knowledge Illusion: Why We Never Think Alone*.

33. Fernbach, P. M., Rogers, T., Fox, C. R., & Sloman, S. A. (2013). Political extremism is supported by an illusion of understanding. *Psychological Science, 24*(6), 939–945; Rozenblit, L., & Keil, F. (2002). The misunderstood limits of folk science: An illusion of explanatory depth. *Cognitive Science, 26*(5), 521–562.

34. Henrich, J. (2015). Culture and social behavior. *Current Opinion in Behavioral Sciences, 3*, 84–89.

35. Lippmann, W. (1922). *Public opinion*. Harcourt, Brace and Company.

36. Hovland, C. I., Lumsdaine, A. A., & Sheffield, F. D. (1949). *Experiments on mass communication*. Princeton University Press.

37. Henrich, J. (2020). *The WEIRDest people in the world: How the West became psychologically peculiar and particularly prosperous*. Penguin UK.

38. Henrich, J. (2016). *The secret of our success*. Princeton University Press.

39. Jones, M. D., McBeth, M. K., & Shanahan, E. A. (2014). Introducing the narrative policy framework. In M. D. Jones, E. A. Shanahan, & M. K. McBeth (Eds.), *The science of stories: Applications of the narrative policy framework in public policy analysis* (pp. 1–25). Palgrave Macmillan.

40. The fairy tales of the fossil fuel industry—and a better climate story | Luisa Neubauer | TED. (n.d.). YouTube. Retrieved September 27, 2014, from https://www.youtube.com/watch?v=wL8X31XWZW8.

41. A beautiful term made popular by late-night television comedian and talk show host Stephen Colbert.

42. Fernbach, P. M., & Bogard, J. P. (2023). Conspiracy theory as individual and group behavior: Observations from the Flat Earth International Conference. *Topics in Cognitive Science, 16*(2), 187–205.

43. Pipes, D. (1997). *Conspiracy: How the paranoid style flourishes and where it comes from.* Free Press. As cited in Whitson, J. A., & Galinsky, A. D. (2008). Lacking control increases illusory pattern perception. *Science, 322*(5898), 115–117. https://www.science.org/doi/10.1126/science.1159845.

Chapter 7

1. Examples like this are discussed at great length in Greene, J. (2014). *Moral tribes: Emotion, reason, and the gap between us and them.* Penguin.

2. This and the subsequent example come from Kahneman, D., & Tversky, A. (1984). Choices, values, and frames. *American Psychologist, 39*(4), 341.

3. This and the subsequent example come from Bazerman, M. H. (1985). Human judgment in managerial decision making. John Wiley and Sons.

4. Mandel, D. R. (2001). Gain-loss framing and choice: Separating outcome formulations from descriptor formulations. *Organizational Behavior and Human Decision Processes, 85*, 56–76.

5. De Martino, B., Kumaran, D., Seymour, B., & Dolan, R. J. (2006). Frames, biases, and rational decision-making in the human brain. *Science, 313*, 684–687.

6. Reyna, V. F., Chick, C. F., Corbin, J. C., & Hsia, A. N. (2014). Developmental reversals in risky decision making: Intelligence agents show larger decision biases than college students. *Psychological Science, 25*(1), 76–84.

7. Kahneman, D., & Tversky, A. (1979). Prospect theory: An analysis of decision under risk. *Econometrica, 47*(2), 279.

8. McNeil, B. J., Pauker, S. G., Sox, H. C., Jr., & Tversky, A. (1982). On the elicitation of preferences for alternative therapies. *New England Journal of Medicine, 306*(21), 1259–1262.

9. Kahneman, D., Knetsch, J. L., & Thaler, R. H. (1990). Experimental tests of the endowment effect and the Coase theorem. *Journal of Political Economy, 98*(6), 1325–1348.

10. Homonoff, T. A. (2018). Can small incentives have large effects? The impact of taxes versus bonuses on disposable bag use. *American Economic Journal: Economic Policy, 10*(4), 177–210.

11. Morewedge, C. K., & Giblin, C. E. (2015). Explanations of the endowment effect: An integrative review. *Trends in Cognitive Sciences, 19*(6), 339–348.

12. Morewedge, C. K., Shu, L. L., Gilbert, D. T., & Wilson, T. D. (2009). Bad riddance or good rubbish? Ownership and not loss aversion causes the endowment effect. *Journal of Experimental Social Psychology*, *45*, 947–951.

13. Yamamoto, S., & Navarro-Martinez, D. (2022). The endowment effect in the future: How time shapes buying and selling prices. *Judgment and Decision Making*, *17*(5), 988–1014.

14. Knutson, B., Wimmer, G. E., Rick, S., Hollon, N. G., Prelec, D., & Loewenstein, G. (2008). Neural antecedents of the endowment effect. *Neuron*, *58*, 814–822.

15. Reviewed in Morewedge & Giblin (2015).

16. Li, Y. J., Kenrick, D. T., Griskevicius, V., & Neuberg, S. L. (2012). Economic decision biases in evolutionary perspective: How mating and self-protection motives alter loss aversion. *Journal of Personality and Social Psychology*, *102*, 550–561.

17. This and the subsequent example come from Kahneman, D., Knetsch, J. L., & Thaler, R. (1986). Fairness as a constraint on profit seeking: Entitlements in the market. *American Economic Review*, 728–741.

18. Tversky, A., & Koehler, D. J. (1994). Support theory: A nonextensional representation of subjective probability. *Psychological Review*, *101*(4), 547.

19. Anderson, E. (1993). *Value in ethics and economics*. Harvard University Press.

20. Slovic, P., & Lichtenstein, S. (1983). Preference reversals: A broader perspective. *American Economic Review*, *73*(4), 596–605.

21. Epley, N., & Gilovich, T. (2001). Putting adjustment back in the anchoring and adjustment heuristic: Differential processing of self-generated and experimenter-provided anchors. *Psychological Science*, *12*(5), 391–396.

22. Slovic, P., Griffin, D., & Tversky, A. (1990). Compatibility effects in judgment and choice. In R. M. Hogarth (Ed.), *Insights in decision making: A tribute to Hillel J. Einhorn* (pp. 5–27). University of Chicago Press.

23. Grether, D. M., & Plott, C. R. (1979). Economic theory of choice and the preference reversal phenomenon. *American Economic Review*, *69*(4), 623.

24. An early paper on the topic is Thaler, R. H. (1990). Anomalies: Saving, fungibility, and mental accounts. *Journal of Economic Perspectives*, *4*(1), 193–205.

25. Milkman, K. L., & Beshears, J. (2009). Mental accounting and small windfalls: Evidence from an online grocer. *Journal of Economic Behavior & Organization*, *71*(2), 384–394.

26. Caro, R. (1974). *The power broker: Robert Moses and the fall of New York*. Knopf.

27. Heath, C. (1995). Escalation and de-escalation of commitment in response to sunk costs: The role of budgeting in mental accounting. *Organizational Behavior and Human Decision Processes, 62*(1), 38–54.

28. This is a lesson that Herbert Simon tried to teach us long ago. See Simon, H. (1957). *Models of man*. John Wiley and Sons.

Chapter 8

1. Schwartz, B. (2004). *The paradox of choice: Why more is less*. Ecco.

2. Roets, A., Schwartz, B., & Guan, Y. (2012). The tyranny of choice: A cross-cultural investigation of maximizing-satisficing effects on well-being. *Judgment and Decision Making, 7*(6), 689–704.

3. Dowling, T. (2012, September 17). Barack Obama's secret weapon? Routine. *Guardian*. https://www.theguardian.com/world/shortcuts/2012/sep/17/barack-obama-secret-weapon-routine.

4. DeScioli, P., & Kurzban, R. (2013). A solution to the mysteries of morality. *Psychological Bulletin, 139*(2), 477.

5. Kruglanski, A. W., Fishbach, A., Woolley, K., Bélanger, J. J., Chernikova, M., Molinario, E., & Pierro, A. (2018). A structural model of intrinsic motivation: On the psychology of means-ends fusion. *Psychological Review, 125*(2), 165.

6. Sansone, C., Sachau, D. A., & Weir, C. (1989). Effects of instruction on intrinsic interest: The importance of context. *Journal of Personality and Social Psychology, 57*(5), 819.

7. Bloom, M. M. (2004). Palestinian suicide bombing: Public support, market share, and outbidding. *Political Science Quarterly, 119*(1), 74.

8. Dawes, R. M., Faust, D., & Meehl, P. E. (1989). Clinical versus actuarial judgment. *Science, 243*(4899), 1668–1674.

9. See Gottfreson, S. D., & Moriarty, L. J. (2006). Clinical versus actuarial judgments in criminal justice decisions: Should one replace the other? *Federal Probation, 70*, 15.

10. Dawes, R. M. (1979). The robust beauty of improper linear models in decision making. *American Psychologist, 34*, 571–582.

11. Payne, J. W., Bettman, J. R., & Johnson, E. J. (1993). *The adaptive decision maker*. Cambridge University Press.

12. Tversky, A. (1972). Elimination by aspects: A theory of choice. *Psychological Review, 79*(4), 281.

13. See Pratt, J. W., Raiffa, H., & Schlaifer, R. (1995). *Introduction to statistical decision theory*. MIT Press.

14. Shafir, E., Simonson, I., & Tversky, A. (1993). Reason-based choice. *Cognition, 49*(1–2), 15–16.

15. Simonson, I. (2014). Vices and virtues of misguided replications: The case of asymmetric dominance. *Journal of Marketing Research, 51*(4), 514–519.

16. For a recent review of the psychology of multiattribute choice, see Bhatia, S., & Stewart, N. (2018). Naturalistic multiattribute choice. *Cognition, 179*, 71–88.

17. Johnson, E. J., Häubl, G., & Keinan, A. (2007). Aspects of endowment: A query theory of value construction. *Journal of Experimental Psychology: Learning, Memory, and Cognition, 33*, 461–474.

18. Franklin, B. (1772, September 19). [Letter from Benjamin Franklin to Joseph Priestley, 1772]. Retrieved from https://founders.archives.gov/documents/Franklin/01-19-02-0200.

19. This idea was developed in Schank, R. C., & Abelson, R. P. (1995). Knowledge and memory: The real story. In R. S. Wyer Jr. (Ed.), *Advances in social cognition, volume VIII* (pp. 1–85). Lawrence Erlbaum Associates.

20. Freling, T. H., Yang, Z., Saini, R., Itani, O. S., & Abualsamh, R. R. (2020). When poignant stories outweigh cold hard facts: A meta-analysis of the anecdotal bias. *Organizational Behavior and Human Decision Processes, 160*, 51–67; Martinez, J. E., Feldman, L. A., Feldman, M. J., & Cikara, M. (2021). Narratives shape cognitive representations of immigrants and immigration-policy preferences. *Psychological Science, 32*(2), 135–152; Kalla, J. L., & Broockman, D. E. (2020). Reducing exclusionary attitudes through interpersonal conversation: Evidence from three field experiments. *American Political Science Review, 114*(2), 410–425.

21. This and the quote above come from Pennington, N., & Hastie, R. (1988). Explanation-based decision making: Effects of memory structure on judgment. *Journal of Experimental Psychology: Learning, Memory, and Cognition, 14*(3), 521.

22. Pennington, N., & Hastie, R. (1993). Reasoning in explanation-based decision making. *Cognition, 49*(1–2), 123–163.

Chapter 9

1. Quoted in DeScioli, P., & Kurzban, R. (2013). A solution to the mysteries of morality. *Psychological Bulletin, 139*(2), 479.

2. See, for example, Niemi, L., & Nichols, S. (in press). Moral decision-making: The value of actions. In B. F. Malle & P. Robbins (Eds.), *The Cambridge handbook of moral psychology*. Cambridge University Press.

3. Greene, J. D., Sommerville, R. B., Nystrom, L. E., Darley, J. M., & Cohen, J. D. (2001). An fMRI investigation of emotional engagement in moral judgment. *Science, 293*(5537), 2105–2108.

4. Sloman, S. A. (1996). The empirical case for two systems of reasoning. *Psychological Bulletin, 119,* 3–22.

5. Park, J., Lee, M., Choi, H., Kwon, Y., Sloman, S., & Halperin, E. (Eds.). (2020). *2019 Annual reports of attitude of Koreans toward peace and reconciliation.* Korea Institute for National Unification.

6. Public Perceptions of the Pay Gap, American Association of University Women Educational Foundation, April 19, 2005.

7. See, for example, Kurzban, R., & Weeden, J. (2014). *The hidden agenda of the political mind: How self-interest shapes our opinions and why we won't admit it.* Princeton University Press.

8. Foot, P. (1967). The problem of abortion and the doctrine of the double effect. *Oxford Review, 5,* 8.

9. Greene, J. D., Sommerville, R. B., Nystrom, L. E., Darley, J. M., & Cohen, J. D. (2001). An fMRI investigation of emotional engagement in moral judgment. *Science, 293*(5537), 2105–2108.

10. Awad, E., Dsouza, S., Shariff, A., Rahwan, I., & Bonnefon, J. F. (2020). Universals and variations in moral decisions made in 42 countries by 70,000 participants. *Proceedings of the National Academy of Sciences, 117*(5), 2332–2337.

11. Greene, J. D., Morelli, S. A., Lowenberg, K., Nystrom, L. E., & Cohen, J. D. (2008). Cognitive load selectively interferes with utilitarian moral judgment. *Cognition, 107,* 1144–1154.

12. Miller, R., & Cushman, F. (2013). Aversive for me, wrong for you: First-person behavioral aversions underlie the moral condemnation of harm. *Social and Personality Psychology Compass, 7*(10), 707–718.

13. Patil, I. (2015). Trait psychopathy and utilitarian moral judgement: The mediating role of action aversion. *Journal of Cognitive Psychology, 27*(3), 349–366.

14. Greene, J. D., Cushman, F. A., Stewart, L. E., Lowenberg, K., Nystrom, L. E., & Cohen, J. D. (2009). Pushing moral buttons: The interaction between personal force and intention in moral judgment. *Cognition, 111,* 364–371.

15. Blair, R. J. R. (1995). A cognitive developmental approach to morality: Investigating the psychopath. *Cognition, 57*(1), 1–29.

16. Kahane, G., Everett, J. A. C., Earp, B. D., Farias, M., & Savulescu, J. (2015). "Utilitarian" judgments in sacrificial moral dilemmas do not reflect impartial concern for the greater good. *Cognition, 134,* 193–209; Bartels, D. M., & Pizarro, D. A. (2011). The mismeasure of morals: Antisocial personality traits predict utilitarian responses to moral dilemmas. *Cognition, 121*(1), 154–161.

17. Kahane, G., Everett, J. A., Earp, B. D., Caviola, L., Faber, N. S., Crockett, M. J., & Savulescu, J. (2018). Beyond sacrificial harm: A two-dimensional model of utilitarian psychology. *Psychological Review, 125*(2), 131.

18. Vsauce. (2017, December 6.) *The trolley problem in real life* [Video]. YouTube. https://www.youtube.com/watch?v=1sl5KJ69qiA&t=1151s.

19. Recounted in Singer, P. (2015). *The most good you can do: How effective altruism is changing ideas about living ethically*. Text Publishing.

20. Everett, J. A., Pizarro, D. A., & Crockett, M. J. (2016). Inference of trustworthiness from intuitive moral judgments. *Journal of Experimental Psychology: General, 145*(6), 772–787.

21. Hidalgo, C. A., Orghian, D., Canals, J. A., De Almeida, F., & Martin, N. (2021). *How humans judge machines*. MIT Press.

Chapter 10

1. Milosavljevic, M., Koch, C., & Rangel, A. (2011). Consumers can make decisions in as little as a third of a second. *Judgment and Decision Making, 6*(6), 520–530.

2. Madl, T., Baars, B. J., & Franklin, S. (2011). The timing of the cognitive cycle. *PloS One, 6*(4), e14803.

3. Wilson, T. D., Lisle, D. J., Schooler, J. W., Hodges, S. D., Klaaren, K. J., & LaFleur, S. J. (1993). Introspecting about reasons can reduce post-choice satisfaction. *Personality and Social Psychology Bulletin, 19*(3), 331–339.

4. Wilson, T. D., & Schooler, J. W. (1991). Thinking too much: Introspection can reduce the quality of preferences and decisions. *Journal of Personality and Social Psychology, 60*(2), 181.

5. Roelofs, J., Rood, L., Meesters, C., te Dorsthorst, V., Bögels, S., Alloy, L. B., & Nolen-Hoeksema, S. (2009). The influence of rumination and distraction on depressed and anxious mood: A prospective examination of the response styles theory in children and adolescents. *European Child & Adolescent Psychiatry, 18*(10), 635–642.

6. Schooler, J. W., & Engstler-Schooler, T. Y. (1990). Verbal overshadowing of visual memories: Some things are better left unsaid. *Cognitive Psychology, 22*(1), 36–71.

7. Schooler, J. W., Ohlsson, S., & Brooks, K. (1993). Thoughts beyond words: When language overshadows insight. *Journal of Experimental Psychology: General, 122*(2), 166.

8. Lloyd-Jones, T. J., & Brown, C. (2008). Verbal overshadowing of multiple face recognition: Effects on remembering and knowing over time. *European Journal of*

Cognitive Psychology, 20(3), 456–477; Wilson, B. M., Seale-Carlisle, T. M., & Mickes, L. (2018). The effects of verbal descriptions on performance in lineups and showups. *Journal of Experimental Psychology: General, 147*(1), 113; Perfect, T. J., Hunt, L. J., & Harris, C. M. (2002). Verbal overshadowing in voice recognition. *Applied Cognitive Psychology, 16*(8), 973–980.

9. Miller, G. A. (1956). The magical number seven, plus or minus two: Some limits on our capacity for processing information. *Psychological Review, 63*(2), 81.

10. Baddeley, A. (2006). Working memory: An overview. *Working Memory and Education,* 1–31.

11. Wilson et al. (1993), p. 331.

12. Grant, N. (2022, July 23). Google fires engineer who claims its A.I. is conscious. *New York Times*. https://www.nytimes.com/2022/07/23/technology/google-engineer-artificial-intelligence.html.

13. Frederick, S. (2005). Cognitive reflection and decision making. *Journal of Economic Perspectives, 19*(4), 25–42.

14. James, W. (1890). *The principles of psychology* (Vol. 1). Henry Holt and Co.

15. Markiewicz, R., Markiewicz-Gospodarek, A., & Dobrowolska, B. (2022). Galvanic skin response features in psychiatry and mental disorders: A narrative review. *International Journal of Environmental Research and Public Health, 19*(20): 13428.

16. Lerner, J. S., Li, Y., Valdesolo, P., & Kassam, K. S. (2015). Emotion and decision making. *Annual Review of Psychology, 66*, 799–823.

17. Damasio, A. (1994). *Descartes' error: Emotion, rationality and the human brain* (p. 352). Putnam.

18. Bechara, A., & Damasio, A. R. (2005). The somatic marker hypothesis: A neural theory of economic decision. *Games and Economic Behavior, 52*(2), 336–372.

19. Basten, U., Biele, G., Heekeren, H. R., & Fiebach, C. J. (2010). How the brain integrates costs and benefits during decision making. *Proceedings of the National Academy of Sciences, 107*(50), 21767–21772; Wunderlich, K., Dayan, P., & Dolan, R. J. (2012). Mapping value based planning and extensively trained choice in the human brain. *Nature Neuroscience, 15*(5), 786–791.

20. Hamid, N., Pretus, C., Atran, S., Crockett, M. J., Ginges, J., Sheikh, H., . . . & Vilarroya, O. (2019). Neuroimaging "will to fight" for sacred values: An empirical case study with supporters of an Al Qaeda associate. *Royal Society Open Science, 6*(6), 181585.

21. Quigley, B. M., & Tedeschi, J. T. (1996). Mediating effects of blame attributions on feelings of anger. *Personality and Social Psychology Bulletin, 22*(12), 1280–1288.

22. Keltner, D., & Lerner, J. S. (2010). Emotion. In S. T. Fiske, D. T. Gilbert, & G. Lindzey (Eds.), *Handbook of social psychology* (pp. 317–352). John Wiley and Sons.

23. Johnson, E. J., & Tversky, A. (1983). Affect, generalization, and the perception of risk. *Journal of Personality and Social Psychology, 45*(1), 20.

24. Schwarz, N., & Clore, G. L. (1983). Mood, misattribution, and judgments of well-being: Informative and directive functions of affective states. *Journal of Personality and Social Psychology, 45*(3), 513.

25. Hirshleifer, D., & Shumway, T. (2003). Good day sunshine: Stock returns and the weather. *Journal of Finance, 58*(3), 1009–1032.

26. Schultz, W., Tremblay, L., & Hollerman, J. R. (1998). Reward prediction in primate basal ganglia and frontal cortex. *Neuropharmacology, 37*(4–5), 421–429.

27. Schultz, W. (2016). Dopamine reward prediction error coding. *Dialogues in Clinical Neuroscience, 18*(1), 23–32.

28. Loewenstein, G. (1999). A visceral account of addiction. In Elster, J., & Skog, O. J. (Eds.), *Getting hooked: Rationality and addiction* (pp. 235–264). Cambridge University Press.

29. DeSteno, D. (2009). Social emotions and intertemporal choice: "Hot" mechanisms for building social and economic capital. *Current Directions in Psychological Science, 18*(5), 280–284.

30. Van Kleef, G. A., De Dreu, C. K., & Manstead, A. S. (2004). The interpersonal effects of emotions in negotiations: A motivated information processing approach. *Journal of Personality and Social Psychology, 87*(4), 510.

Chapter 11

1. This is a summary of Grann, D. (2009, August 31). Trial by fire. *New Yorker*. https://www.newyorker.com/magazine/2009/09/07/trial-by-fire.

2. Fridman, A., Gershon, R., & Gneezy, A. (2021). COVID-19 and vaccine hesitancy: A longitudinal study. *PloS One, 16*(4), e0250123.

3. Hornsey, M. J., Finlayson, M., Chatwood, G., & Begeny, C. T. (2020). Donald Trump and vaccination: The effect of political identity, conspiracist ideation and presidential tweets on vaccine hesitancy. *Journal of Experimental Social Psychology, 88*, 103947.

4. Summers, J. (2020, October 2). Timeline: How Trump has downplayed the coronavirus pandemic. NPR. https://www.npr.org/sections/latest-updates-trump-covid-19-results/2020/10/02/919432383/how-trump-has-downplayed-the-coronavirus-pandemic.

5. Fridman, Gershon, & Gneezy (2021).

6. Geana, M. V., Rabb, N., & Sloman, S. (2021). Walking the party line: The growing role of political ideology in shaping health behavior in the United States. *SSM-Population Health, 16*, 100950.

7. Fullerton, M. K., Rabb, N., Mamidipaka, S., Ungar, L., & Sloman, S. A. (2022). Evidence against risk as a motivating driver of COVID-19 preventive behaviors in the United States. *Journal of Health Psychology, 27*(9), 2129–2146.

8. Gozdzielewska, L., Kilpatrick, C., Reilly, J., Stewart, S., Butcher. J., Kalule, A., . . . & Price, L. (2022). The effectiveness of hand hygiene interventions for preventing community transmission or acquisition of novel coronavirus or influenza infections: A systematic review. *BMC Public Health, 22*(1), 1283.

9. Willingham, D. T. (2008). Critical thinking: Why is it so hard to teach? *Arts Education Policy Review, 109*(4), 21–32.

10. Galinsky, A. D., & Ku, G. (2004). The effects of perspective-taking on prejudice: The moderating role of self-evaluation. *Personality and Social Psychology Bulletin, 30*(5), 594–604.

11. Lord, C. G., Lepper, M. R., & Preston, E. (1984). Considering the opposite: A corrective strategy for social judgment. *Journal of Personality and Social Psychology, 47*(6), 1231.

12. Stanovich, K. E., & Toplak, M. E. (2023). Actively open-minded thinking and its measurement. *Journal of Intelligence, 11*(2), 27.

13. Soll, J. B., Milkman, K. L., & Payne, J. W. (2015). A user's guide to debiasing. In K. Gideon & G. Wu (Eds.), *The Wiley Blackwell handbook of judgment and decision making* (Vol. 2, pp. 924–951). Wiley-Blackwell.

14. Pacini, R., & Epstein, S. (1999). The relation of rational and experiential information processing styles to personality, basic beliefs, and the ratio-bias phenomenon. *Journal of Personality and Social Psychology, 76*(6), 972.

15. Ludolph, R., & Schulz, P. J. (2018). Debiasing health-related judgments and decision making: A systematic review. *Medical Decision Making, 38*(1), 3–13.

16. Willingham (2008), p. 21.

17. Soll, Milkman, & Payne (2015).

18. Larrick, R. P. (2004). Debiasing. In D. J. Koehler & N. Harvey (Eds.), *The Wiley Blackwell handbook of judgment and decision making* (Vol. 1, pp. 316–338). Blackwell Publishing.

19. Lazer, D., Baum, M., Benkler, Y., Berinsky, A., Greenhill, K., . . . Sloman, S., . . . & Zittrain, J. (2018). The science of fake news. *Science, 359*(6380), 1094–1096.

20. Pennycook, G., & Rand, D. G. (2019). Lazy, not biased: Susceptibility to partisan fake news is better explained by lack of reasoning than by motivated reasoning. *Cognition, 188*, 39–50.

21. Nieweglowska, M., Stellato, C., & Sloman, S. A. (2023). Deepfakes: Vehicles for radicalization, not persuasion. *Current Directions in Psychological Science, 32*(3), 236–241.

22. Larrick (2004).

23. Rappeport, A. (2016, September 8). "What is Aleppo?" Gary Johnson asks, in an interview stumble. *New York Times*. https://www.nytimes.com/2016/09/09/us/politics/gary-johnson-aleppo.html.

24. Rozenblit, L., & Keil, F. (2002). The misunderstood limits of folk science: An illusion of explanatory depth. *Cognitive Science, 26*(5), 521–562.

25. Sloman, S., & Fernbach, P. (2017). *The knowledge illusion: Why we never think alone*. Riverside Press.

26. Ripley, A. (2018, June 27). Complicating the narratives. *The Whole Story*. https://thewholestory.solutionsjournalism.org/complicating-the-narratives-b91ea06ddf63.

Chapter 12

1. Chauvet, L., Gubert, F., & Mesplé-Somps, S. (2016). Do migrants adopt new political attitudes from abroad? Evidence using a multi-sited exit-poll survey during the 2013 Malian elections. *Comparative Migration Studies, 4*, 1–31.

2. Loewenstein, G. (1999). A visceral account of addiction. In J. Elster & O. J. Skog (Eds.), *Getting hooked: Rationality and addiction* (pp. 235–264). Cambridge University Press.

3. Gigerenzer, G. (2004). Dread risk, September 11, and fatal traffic accidents. *Psychological Science, 15*(4), 286–287.

4. Sloman, S. A., & Vives, M. L. (2022). Is political extremism supported by an illusion of understanding? *Cognition, 225*, 105146.

5. Missmahl, I. (2018). Value-based counselling: Reflections on fourteen years of psychosocial support in Afghanistan. *Intervention, 16*(3), 256–260.

6. Missmahl (2018), p. 257.

7. Lukianoff, G., & Haidt, J. (2018). *The coddling of the American mind: How good intentions and bad ideas are setting up a generation for failure* (p. 30). Penguin.

Index